Enrico Rizzo

Simulations for the optimization of High Temperature Superconductor current leads for nuclear fusion applications

Karlsruher Schriftenreihe zur Supraleitung Band 014

HERAUSGEBER

Prof. Dr.-Ing. M. Noe
Prof. Dr. rer. nat. M. Siegel

Eine Übersicht über alle bisher in dieser Schriftenreihe erschienene Bände finden Sie am Ende des Buches.

Simulations for the optimization of High Temperature Superconductor current leads for nuclear fusion applications

by
Enrico Rizzo

Dissertation, Karlsruher Institut für Technologie (KIT)
Fakultät für Elektrotechnik und Informationstechnik, 2013
Hauptreferent: Prof. Dr.-Ing. Mathias Noe
Korreferent: Prof. Dr.-Ing. Roberto Zanino

Impressum

Karlsruher Institut für Technologie (KIT)
KIT Scientific Publishing
Straße am Forum 2
D-76131 Karlsruhe

KIT Scientific Publishing is a registered trademark of Karlsruhe Institute of Technology. Reprint using the book cover is not allowed.

www.ksp.kit.edu

Print on Demand 2014

ISSN 1869-1765
ISBN 978-3-7315-0132-9
DOI: 10.5445/KSP/1000037132

Simulations for the optimization of High Temperature Superconductor current leads for nuclear fusion applications

Zur Erlangung des akademischen Grades eines

DOKTOR-INGENIEURS

von der Fakultät für

Elektrotechnik und Informationstechnik

des Karlsruher Instituts für Technologie (KIT)

genehmigte

DISSERTATION

von

M. Sc. Enrico Rizzo

geboren in: Asti (Italien)

Tag der mündlichen Prüfung: 29. Oktober 2013

Hauptreferent: Prof. Dr.-Ing. Mathias Noe

Korreferent: Prof. Dr.-Ing. Roberto Zanino (Politecnico di Torino)

Acknowledgement

This work could not have been realized without the supervision and professional support of Dr. Reinhard Heller (KIT, ITEP), of Prof. Dr. Mathias Noe (KIT, ITEP), of Prof. Dr. Roberto Zanino (Politecnico di Torino, DENERG) and, of course, of tireless Dr. Laura Savoldi Richard (Politecnico di Torino, DENERG).

To them and to all others who have shared a piece of their life with me, as well as to my family, I wish I will be able to show every day my boundless gratitude.

Karlsruhe, August 14th 2013

Enrico Rizzo

Kurzfassung

Die Kernfusion, basierend auf dem magnetischen Plasmaeinschluss, ist eine erfolgversprechende Energiequelle für die Zukunft der Menschheit. Um das Plasma einzuschließen, sind starke magnetische Felder nötig, die mit supraleitenden Fusionsmagneten erzeugt werden. Die Betriebstemperatur heutiger Fusionsmagnete liegt bei etwa 4,5 K, d.h. im superkritischen Bereich von Helium. Daher werden Fusionsmagnete innerhalb eines Kryostaten betrieben und durch Stromzuführungen an eine Raumtemperaturstromquelle angeschlossen.

Da die Stromzuführungen aus leitfähigen Materialien hergestellt werden (z. B. Kupfer oder Aluminium), kommt es zu einem Wärmeeintrag in den Kryostaten. Dieser Wärmeeintrag besteht aus zwei Anteilen, dem durch den Temperaturgradienten hervorgerufenen Wärmestrom und die Joule'sche Wärme. Daher müssen die Stromzuführungen gekühlt werden, beispielweise durch Heliumdampf, der aus einem Heliumbad am kalten Ende der Stromzuführungen heraussiedet.

Die Kühlleistung, die bereitgestellt werden muss um die Stromzuführungen zu betreiben, kann stark reduziert werden, wenn statt konventionellen Leitern im Temperaturbereich 4,5 – 70 K Hochtemperatur-Supraleiter verwendet werden. In diesem Fall wird keine Joule'sche Wärme erzeugt. Eine solche Stromzuführung wird als Hochtemperatur-Supraleiter (*HTS*)-Stromzuführung bezeichnet und besteht prinzipiell aus:

- einem *HTS* Modul, das den Strom über dem Temperaturbereich 4,5 – 70 K zuführt, und
- einem konventionellen Leiter, als Wärmetauscher bezeichnet, der den Strom über dem Temperaturbereich 70 – 300 K zuführt.

Beide Komponenten werden in Reihe geschaltet. Wie in einer konventionellen Stromzuführung muss der Wärmetauscher von gasförmigem Helium gekühlt werden. Daher sollte eine möglichst effektive Wärmeübertragung zum Kühlmedium vorhanden sein.

Das Karlsruher Institut für Technologie, *KIT*, hat während des vergangenen Jahrzehntes gezeigt, dass *HTS* Stromzuführungen, welche *BSCCO* Bänder als Hochtemperatursupraleiter verwenden, eine erfolgversprechende Alternative zu konventionellen Stromzuführungen für Fusionsmagnete sind.

In dieser Arbeit werden Methoden und Verfahren entwickelt, um das Design von *HTS* Stromzuführungen für Fusionsmagnete zu optimieren.

Die nötigen Grundlagen, um dieses Ziel zu erreichen, werden in den Kapiteln I, II und III dargelegt.

In den Kapiteln IV und V wird die Fluidmechanik von Helium untersucht, welches die Wärmetauscher innerhalb der *HTS*-Stromzuführungen kühlt. Solche Wärmetauscher sind durch eine besondere, mäanderförmige Geometrie charakterisiert, auch *meander flow* Geometrie genannt, und werden in den *HTS* Stromzuführungen für das *LHC* am *CERN* sowie in den für den *W7-X* Stellarator verwendet. Außerdem werden die Stromzuführungen für die Tokamaks *JT-60SA* und *ITER* die gleiche Art von Wärmetauscher haben.

Die Untersuchung wird mittels einer numerischen Fluidmechanik-Analyse durchgeführt. Da die Geometrie des Wärmetauschers periodisch ist und die Variation der Heliumeigenschaften innerhalb einer Periode der Geometrie begrenzt ist, kann die numerische Analyse in einer einzelnen Periode durchgeführt werden. Der Einfluss der Heliumeigenschaften und des Heliummassenstromes, der in dieser Periode fließt, sowie der Geometrieparameter auf die Fluidmechanik wird systematisch untersucht. Hierzu wurden folgende Untersuchungen durchgeführt:

- Bestimmung des hydraulischen Durchmessers d_{h} und des charakteristischen Heliumstromquerschnitts A_{He},
- Reynolds-Zahl basierte Begrenzung der laminaren und turbulenten Strömungsbereiche,
- Herleitung von Wechselbeziehungen für den Druckabfall-Koeffizienten und die Nusselt-Zahl in beiden Strömungsbereichen, die von der Reynolds-Zahl und weiteren dimensionslosen Verhältnissen zwischen geometrischen Parametern abhängig sind.

Diese Ergebnisse erlauben eine präzise und effektive Auslegung des Wärmetauschers sowie die Optimierung der Stromzuführung.

In Kapitel VI werden numerische Techniken für das Design von *HTS* Modulen entwickelt, die auf der Finite-Elemente-Methode basieren. Diese Techniken dienen dazu, das Verhalten des *HTS* Moduls im Normal- sowie im Fehler-Betrieb vorauszuberechnen, so dass optimierte Lösungen zum Aufbau gefunden werden können. Diese Techniken werden mit den experimentellen Ergebnissen des *ITER* 70 kA Demonstrators und der *W7-X* Stromzuführungen validiert.

In Kapitel VII werden die Korrelationen aus Kapitel V angewendet. Es wird eine stationäre Analyse der drei *ITER* Stromzuführungstypen durchgeführt und mit den Spezifikationen verglichen. Den Ergebnissen dieser Analyse entsprechend werden die *ITER* Stromzuführungen die Anforderungen erfüllen können.

Table of content

1 Introduction and motivation **1**

2 Scope and main features of High Temperature Superconductor current leads **5**

2.1 Scientific and technological application ... 5

2.2 Current lead's technology ... 6

2.2.1 Resistive current leads ... 6

2.3 HTS current leads ... 12

2.3.1 Room temperature terminal and heat exchanger ... 12

2.3.2 HTS module ... 15

2.3.3 Low temperature connection ... 16

2.3.4 Joints and other interconnections ... 16

3 Optimization of High Temperature Superconductor current leads **19**

3.1 Optimization procedure ... 19

3.1.1 Mathematical formulation for the thermal optimization of HTS current leads ... 19

3.1.2 Thermal-hydraulics of the coolant ... 23

3.2 Design tool for the HTS module ... 24

4 CtFD analysis of the helium flow in the meander flow heat exchanger **27**

4.1 Meander flow heat exchanger ... 27

4.2 CtFD analysis based on the periodic modelling ... 28

4.2.1 Scope of the CtFD analysis ... 32

4.2.2 Ranges for the CtFD analysis ... 32

4.2.3 Development of the periodic modelling for the CtFD analysis 35

4.2.4 Strategy for the CtFD analysis with the periodic modelling ... 39

4.2.5 Flow regime and physical models ... 45

4.2.6 Software, numerical method, mesh generation and solution algorithm ... 52

4.3 Validation of the CtFD analysis based on the periodic modelling...... 58
4.4 Summary...... 62

5 Correlations for the helium thermal-fluid dynamics in the meander flow geometry 65
5.1 Characteristic geometrical quantities of the meander flow geometry 65
5.1.1 Characteristic helium flow cross section, A_{He}...... 66
5.1.2 Hydraulic diameter, d_{h}...... 72
5.2 Flow regime inside the meander flow geometry 77
5.2.1 Definition of the turbulent regime...... 77
5.2.2 Definition of the laminar regime 80
5.3 Correlation for the pressure drop coefficient, ζ...... 82
5.3.1 Correlation for ζ in the turbulent regime...... 83
5.3.2 Correlation ζ in the laminar regime 88
5.4 Correlation for the Nusselt number, Nu 94
5.4.1 Correlation for Nu in the turbulent regime...... 95
5.4.2 Correlation for Nu in the laminar regime 101
5.5 Summary...... 106

6 Numerical modelling of the HTS module 111
6.1 Design issues of the HTS module 111
6.1.1 Computational domains for the modelling of the HTS module 114
6.1.2 Mathematical model for the analysis...... 116
6.1.3 Software, numerical method and mesh 120
6.1.4 Analysis with the 2-D axis-symmetric and 3-D reduced models...... 120
6.2 Validation against experimental results 121
6.2.1 Steady-state analysis...... 122
6.2.2 Time-dependent analysis...... 125
6.3 Summary...... 129

7 Predictive analysis of ITER HTS current leads 131
7.1 ITER HTS current leads...... 131
7.2 Scope and methodology of the analysis 134

7.2.1 CURLEAD code ... 135
7.2.2 1-D models of ITER HTS current leads ... 136
7.2.3 Procedure of the 1-D analysis ... 139
7.3 Results of the analysis ... 140
7.3.1 Influence of the contact resistance ... 140
7.3.2 Heat load at the cold end of the current leads ... 144
7.3.3 Convective heat transfer ... 145
7.4 Summary ... 149

8 Conclusion and perspectives **151**

Appendix A HTS conductors of technical interest **157**
Bismuth-strontium-calcium-copper-oxide, BSCCO ... 157
Rare earth-bismuth-copper-oxide, *RE*BCO ... 160

Appendix B Forthcoming nuclear fusion experiments **163**
Tokamak ... 163
Stellarator ... 171

Appendix C Material properties **177**
Helium ... 177
Copper ... 179
Stainless steel ... 181
BSCCO stacks ... 181

Appendix D Grid independence analysis **185**
CtFD periodic modelling ... 185
HTS module modelling ... 188

Table of symbols **191**

Table of abbreviations and indexes **193**

Literature **195**

1 Introduction and motivation

Every year, the International Energy Agency, IEA, publishes a detailed and exhaustive outlook [WEO12] on the worldwide energy consume and associated issues. Additionally, several predictive scenarios about the evolution of energy consumption over the next 20 years are updated on a yearly base. In the 2012 edition of the outlook it is shown that all scenarios on global energy trends predict that [WEO12, p. 50]:

- the world energy needs will rise,
- the dynamics of the energy markets will be determined by emerging economies,
- fossil fuels (oil, coal and natural gas) will continue to meet most of the world's energy needs,
- the universal energy access to the world's poor will persistently fail.

The rise in energy demand is driven both by the world population increment and by constantly enlarging fractions of developing economies' population pursuing for west-world-like life-standards. Among the several and severe consequences of these scenarios, the emission of greenhouse gases and the consequent global warming are worldwide of concern: within the end of this century, the average temperature on the Earth is expected to increase between 2°C and 5.6°C with respect to the pre-industrial period because of the anthropic activity [WEO12, p. 52]. Though, fossil fuels will still constitute the kern of primary energy source in twenty years from now, despite of the evolution scenario considered. This means that even if very optimistic assumptions are made about the development, support and adoption of present sustainable sources of primary energy, the mid-term future depends on burning fossil fuels. Moreover, a likely large fraction of the consumed fossil fuels could be coal [WEO12, p. 50].

Against this background, the search for innovative energy sources assumes a central role. In this respect, the next few years are a crucial gateway for finding out if the nuclear fusion based on the magnetic confinement can become a long-term breakthrough and sustainable energy technology (in 50-70 years from now). Hopes of a large part of the scientific community are mainly pinned on the success of the

International Thermonuclear Experimental Reactor, ITER, and on a few other experimental devices (see Appendix B for more information).

Nuclear fusion based on the magnetic confinement[1] aims at exploiting the energy released by fusion reactions between light nuclei (i.e. deuterium and tritium) occurring inside a fully ionized plasma. The goal is to convert this energy into electricity. Nuclear fusion would have significant advantages with respect to both fossil fuels and to nuclear fission energy. Indeed, nuclear fusion energy has near zero carbon dioxide emission and fuel resources are distributed more homogeneously worldwide; moreover, nuclear waste would consist of activated materials both in a smaller quantity and with a much shorter life time than the waste from nuclear fission power plants. In terms of operating conditions, it would provide the base load of the electricity demand; therefore the nuclear fusion energy would suit a supply system with intensive exploitation of renewable energy. On the other hand, a nuclear fusion reactor is a far more complicated machine than any other in the energy technology. The physical and technological challenge is presently at the limit of the human possibilities. The development of a commercial-sized nuclear fusion reactor has required an international joint effort, which has been focused towards the above mentioned ITER project. This project is presently also the most expensive scientific experiment ever built: the price was estimated to be close to 16 billion euros (investment and operation, 2008) [EU10].

Magnetic fields in the order of several tesla are required to realize the confinement of the plasma; powerful electro-magnets are therefore needed. Nevertheless, normal conducting magnets are not suited for the purpose of commercial nuclear fusion power; indeed, the cooling power required for operating a resistive magnet system would largely overcome the output power of the reactor. The alternative is to use a superconducting magnet system. Although the use of superconductors represents a fundamental step forward from research to commercial fusion reactor, it also complicates the design of the machine. In the first generation of superconducting

[1] As mentioned here, this work deals with nuclear fusion energy based on the magnetic confinement. In the prosecution, the author always refers to this kind of process whether not explicitly mentioned.

magnet for fusion reactors, magnets are wound out of Low Temperature Superconductor cables (see, for instance, [Bar13, Chap. II]). In order to operate properly, they must be kept at 4.5 K. For this reason, they are cooled by supercritical helium and housed inside a huge cryostat.

Although the cryostat is a physical barrier aimed at preserving conditions others than the environmental ones, a nuclear fusion reactor cannot be a closed system; the inner part of the cryostat has to be connected to the outside and each opening is critical for maintaining the reactor operative conditions.

One kind of connections across the cryostat is needed to power the superconducting magnet system. The electrical current required to energize the magnets inside the cryostat at 4.5 K must be transported from a room temperature power supply. The devices that realize the connection between the room temperature power supply and the superconducting magnet system are the current leads.

Intuitively, current leads are delicate components of a nuclear fusion reactor: since they have to transport the electrical current, they must be made of conducting material; nevertheless, good electrical conductors are, in turn, good heat conductors. Current leads are therefore favorable gates across which heat flows inside the cryostat due to the temperature gradient and to resistive losses. Moreover, since the electrical current to be transported is rated at tens of kilo-ampere, current leads must be cooled. A central goal of the design of current leads is the reduction of cooling power required to operate them.

A significant reduction of the cooling power can be achieved by limiting the resistive losses at the cold side of the current leads. This can be done adopting current leads partly made of High Temperature Superconductor, HTS, and partly of normal conductor. The HTS part of such a current lead, named HTS module, transports the electrical current approximately in the temperature range 4.5 – 70 K, whereas the normal conductor in the range 70 – 300 K. The normal conductive part must be cooled by means of nitrogen vapour or gaseous helium. For this reason, it is normally provided with extended cooling surfaces to enhance the heat transfer and it is referred to as heat exchanger.

The Karlsruhe Institute of Technology, KIT, has demonstrated for the first time the feasibility of HTS current lead for fusion applications [HFK03, HDD05], it is building the HTS current leads for the tokamak JT-60SA and has recently delivered the HTS current leads for the stellarator W7-X [HDF11a, FHK09] (see Appendix B for details on these experimental nuclear fusion reactors). The HTS current leads for JT-60SA and W7-X mount a so-called meander flow heat exchanger, which is cooled by gaseous helium; the HTS module basically consists of a hollow stainless steel cylinder provided with copper ends and on which HTS BSCCO tapes are arranged into panels.

Although power consumption of HTS current leads is significantly lower than that of current leads made out of normal conductor, the design aims at minimizing it along with the heat load inside the cryostat still allowing the HTS current lead to be operated properly. The procedure is named *optimization* and it is relevant for the operation of the entire nuclear fusion reactor.

In this work, techniques are developed with the purpose of designing and optimizing HTS current leads for fusion applications. These techniques involve the simulation, or numerical modelling, of two relevant aspects: the helium flow in the meander flow heat exchanger and the operation of the HTS module.

The work is organized as follows: Chapter II and III offer a general overview on conventional and superconducting current leads and on the mathematical model for the optimization. In Chapters IV and V, a previously introduced and validated (by this author) computational technique for the thermal-fluid mechanics is applied to the systematic analysis of the helium cooling the meander flow heat exchanger for HTS current leads. In Chapter VI, computational models for the HTS module of the HTS current leads are developed and validated. In Chapter VII, the results of the helium thermal fluid mechanics analysis presented in Chapter IV and V are applied to a predictive analysis of the HTS current leads for the superconducting magnets system of ITER. Conclusion and perspectives of this work are presented in Chapter VIII.

2 Scope and main features of High Temperature Superconductor current leads

This introductive Chapter provides the reader with the background about the scope and the main feature of HTS current leads. A brief excursus on the current lead technology over the last fifty years clarifies the advantages behind the adoption of HTS current leads. In conclusion, the main features of HTS current leads are presented, with particular emphasis on the KIT design of the HTS current leads for nuclear fusion application.

2.1 Scientific and technological application

Superconducting devices must be kept at cryogenic temperature (typically $T < 77$ K) in order to operate properly. For this reason, they are placed into a cryostat, which maintains the device at the right conditions and separates it from the external environment. As it is well known from thermodynamics, the cooling power to be provided and the associated costs to be covered for the cool-down and the maintenance of appropriate operative conditions for such a superconducting device depend on the lower temperature to be achieved, on the heat load due to the operation of the device and on the undesired heat losses of the cryostat. In particular, the latter contribution could be set to zero only with a perfectly insulating cryostat, completely shielding its inside from the external contribution. Nevertheless, the electrical current has to be carried somehow into the cryostat. For this reason, it is necessary to connect an electrical power supply at room temperature to the device within the cryostat. The devices that realize this connection are called current leads.

The presence of current leads increases the heat leak inside the cryostat and therefore the cooling power that is required. The most important goal of the design of current leads is to achieve an *as low as possible* heat leak inside the cryostat in normal operation, still safely and reliably transporting the electrical current demanded by the device. This procedure is referred to as *optimization* of current leads. In case of failure, current leads have also to fulfil some safe criteria in order to allow the discharge and a safe recovery of the superconducting device, without disruptive damages.

2.2 Current lead's technology

As stated above, a current lead realizes the connection between a room temperature electrical power supply and a, typically superconducting, device at cryogenic temperature. It has to transport a specific electrical current along the temperature gradient from room temperature to cryostat's temperature and, at the same time, to limit the heat leak inside the cryostat. An example of the circuit connecting a superconducting magnet inside the cryostat via current leads is shown in Fig. 2.1 a).

The first option to build current leads has been widely applied throughout the last fifty years and consists in using normal conducting materials, e.g. copper or aluminium[2]. The goal of the design is to shape the current lead in order to fulfil the above-mentioned requirements. The literature dedicated to this topic is relatively large, but the author thinks that a good and simple explanation for pointing out the main scope of the current lead technology has been provided in [BC76]:

"The design of a pair of cryogenic current leads is a trade-off between carrying a large current with little resistive dissipation and limiting the conduction heat leak housed by the leads."

2.2.1 Resistive[3] current leads

Two physical phenomena have to be taken into account: the heat conduction due to the temperature gradient between the warm and the cold end of the current lead and the Joule heat due to the transported electrical current. Normal conducting materials such as copper or aluminium obey fairly well the well-known Wiedemann-Franz law [WF1853], at least if the temperature is larger than about T = 40 K. According to the

[2] In the prosecution, the focus is on copper because of the relevance for this work.

[3] Current leads made of normal conducting materials can be referred to as resistive, conventional or also normal current leads. The nomenclature is not a major issue; so long the kind of current lead being discussed is clear. In this work, current leads made of a normal conductor will be referred to as resistive current leads. The adjective resistive will be also used in the following to indicate the normal conducting part of HTS current leads.

Wiedemann-Franz law, the product of the two transport properties associated with the above-mentioned phenomena, namely the thermal conductivity $\lambda_{Cu}(T)$ and the electrical resistivity $\rho_{Cu}(T)$, is proportional to the temperature T:

$$\lambda_{Cu}(T) \cdot \rho_{Cu}(T) = L_o \cdot T, \tag{2.1}$$

where L_o is the proportionality constant named Lorenz number, with L_o = 2.44·10-8 W·Ω/K.

Resistive conduction-cooled current leads

A resistive conduction-cooled current lead is made of normal conducting material and operates between room temperature, T_W, and a cold temperature T_C, as shown in Fig 2.1 b). An exhaustive, analytical treatment of the design and optimization of a conventional conduction-cooled current lead can be found in [BFS75].

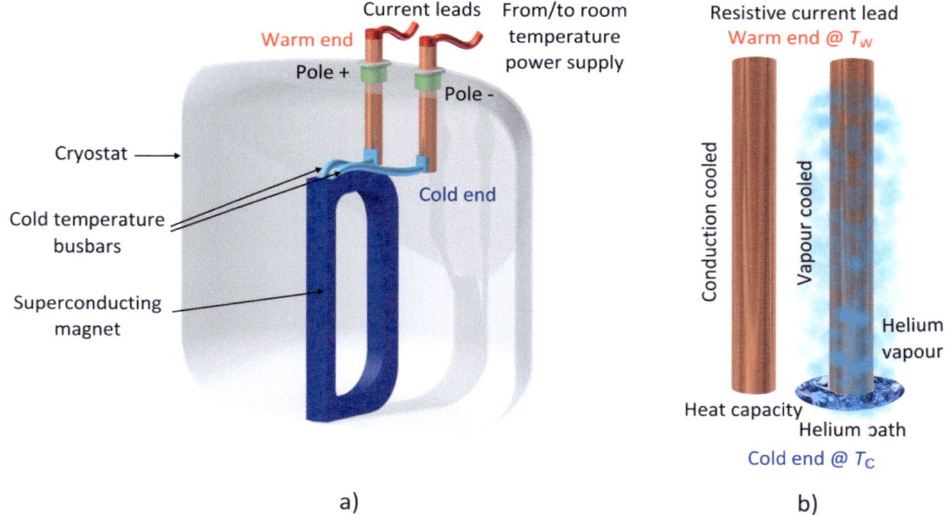

Fig. 2.1: a) Current leads connecting a superconducting magnet inside the cryostat to a room temperature power supply;

b) Cooling of resistive current leads: in the conduction cooled case, the cold end is kept at the cold temperature; in the vapour cooled case, the cold end is in contact with a helium bath; helium is partially vaporized due to the inflowing heat flux.

For the purpose of this section it is worth recalling that the heat conducted, $\dot{Q}_{\text{cond}}$, along a portion of conductor of length l and cross section A_{Cu} and the heat power released within this portion by the Joule heating, $\dot{Q}_{\text{Joule}}$, are respectively proportional to:

$$\dot{Q}_{\text{cond}} \propto \frac{A_{\text{Cu}}}{l}, \tag{2.2}$$

and

$$\dot{Q}_{\text{Joule}} \propto \frac{l}{A_{\text{Cu}}}. \tag{2.3}$$

The heat conducted and the Joule heating $\dot{Q}_{\text{cond}}$ and $\dot{Q}_{\text{Joule}}$ behave opposite with respect to the geometry of the conductor; for a specific value of the transported current I, it is possible to shape the ratio l/A_{cu} of the conductor in order to balance the contribution of the heat conduction and the Joule heating: the heat load associated with the current lead can be therefore minimized. It is worthwhile noting that, as long as the material used as conductor behaves according to the Wiedemann-Franz law, the design of a current lead does not depend on the material itself. Resistive conduction-cooled current leads are used to transport electrical currents up to a few hundred amperes ($I = 100 - 200$ A) in particular when [Bal04]:

- the connection between the external electrical power source and the device inside the cryostat cannot be realized along a preferable straight path,
- the active cooling of the current lead is not feasible.

Resistive vapour-cooled current leads

An alternative to a convection-cooled current lead is a resistive vapour-cooled current lead. Like the previous kind, it is made of normal conductor and it is operating between the temperatures T_{W} and T_{C}. Unlike a resistive conduction-cooled current lead, the cold end of a vapour-cooled current lead is in contact with a helium- ($T_{\text{C}} = 4.2$ K) or, possibly, liquid nitrogen bath ($T_{\text{C}} = 77$ K). The heat leak at the cold end of the current lead into the bath vaporizes a fraction of the coolant. The vapour flows along the current lead towards its warm end and cools the current lead down. A vapour-cooled current lead is therefore analogous to a heat exchanger: resistive thermal losses as well as the conducted heat are carried away by the vapour.

An exemplarily view of a vapour cooled current lead is shown in Fig. 2.1 b).

Procedures for designing and optimizing this kind of current lead have been presented in several publications, as shown in [BFS75] and [Buy85]. A widely adopted analytical treatment has been summarized in [Wils89, p. 256] for a resistive helium vapour-cooled current lead. The objective of the design and optimization are the same as for a resistive conduction-cooled current lead, i.e. the minimization of the cooling power required to safely and reliably operate the lead. The contributions of the heat conduction and of the Joule heat have still to be balanced, but in this case also the heat transfer to the vapour has to be considered. The goal is to minimize the quantity of gas evaporated at the cold end of the current lead, while maximizing the enthalpy change of the vapour along the lead itself [Wils89, p. 256].

This kind of current lead can be used:

- if the electrical current to be transported makes the use of conduction-cooled current leads unpractical,
- if the coolant, i.e. liquid helium or nitrogen, is available.

Performance of the resistive current leads

Analytical solutions for optimizing resistive current leads can be sought for both conduction- and vapour-cooled cases. For a conductor obeying the Wiedemann-Franz law and operating between T_W = 300 K and T_C = 4.2 K, the theoretical, minimum heat leak $\dot{Q}_{C,o}$ of a resistive conduction-cooled current lead is $\dot{Q}_{C,o}$ = 47 mW/A [Wils89, p. 260]; the theoretical, minimum heat leak of a resistive vapour-cooled current lead is $\dot{Q}_{C,o}$ = 1.04 mW/A [Wils89, p. 260]. From this point of view, it is clear that a vapour-cooled current lead is, in principle, preferable with respect to a conduction-cooled current lead; on the other hand, the two kinds of current lead differ in terms of applicability; therefore, the choice among them involves further considerations than just the minimum heat leak.

The cooling power, P_t, required to operate both kinds of current lead has to be provided at the cold end of the current leads themselves. For a resistive conduction-cooled current lead, the cooling power is required to operate the refrigerator keeping the cold end of the lead at the cold temperature T_C; for a resistive vapour-cooled current lead, the cooling power is required to liquefy the helium (or the nitrogen). For both cases, the cooling power P_t depends on the specific heat leak $\dot{Q}_{C,o}$, on the inverse of the Carnot efficiency η_c and on a factor f which is always larger than one, as shown in Eq. 2.4:

$$P_t = \dot{Q}_{C,o} \cdot \frac{1}{\eta_c} \cdot f. \tag{2.4}$$

The Carnot efficiency η_c is the maximum, theoretically achievable efficiency of a thermodynamic machine (operating a Carnot cycle, *Ed.*) and depends on the highest and lowest temperatures T_W and T_C, respectively. The expression for the Carnot efficiency is shown in Eq. 2.5. The factor f accounts for the losses and the irreversibility typical of a real thermodynamic machine, e.g. refrigerator or gas liquefier.

$$\eta_c = 1 - \frac{T_C}{T_W}. \tag{2.5}$$

The reduction of the cooling power P_t required to operate a resistive current lead has been pursued in several ways. According to Eq. 2.5, it is clear that the lower temperature T_C highly influence the cooling power; indeed, it is convenient to reduce the heat leak at the lower temperature. In more detail, at low temperature it is convenient to have as low as possible thermal losses due to the Joule heat, $\dot{Q}_{Joule}$. A possible solution is to design a resistive current lead with a variable cross section A_{Cu}. In this way, the Joule heat $\dot{Q}_{Joule}$ could be reduced at the cold end of the current lead by increasing the cross section A_{Cu}; however, the benefit of a reduced Joule heat would be partly cancelled by the inevitably larger conducted heat $\dot{Q}_{cond}$, as shown in Eq. 2.2. This solution could be also supported by a multi-stage cooling approach of the resistive current lead [Hil77, CVS98]. Unfortunately, although theoretically viable, these solutions are technically challenging. For instance, the manufacturing of a resistive current lead with varying cross section is possible, but it is more complicated than the case with constant cross section; also its montage inside the machine could not be so straightforward. Furthermore, for practical applications the multi-stage cooling consists typically of two/three cooling stages; it would be therefore more advantageous than a single cooling station, but still not the theoretical infinite cooling stations approximation which can maximize the performance of the current lead.

The discovery of High Temperature Superconductors, HTS, and the relatively fast commercialization of these materials (see Appendix A) provided the possibility to transport the electrical current at a temperature $T \leq 70$ K with this kind of superconductors. In [MRV89], the feasibility of this kind of current lead was first demonstrated with a theoretical approach. At the same time, it was demonstrated in [Mum89] that the use of the HTS conductor YBCO instead of phosphorus deoxidised copper in the temperature range 5 – 77 K in a current lead operating between room temperature and liquid helium temperature reduces the heat leak $\dot{Q}_{C,o}$ up to a factor six. Soon after, HTS current leads began to be chosen for worldwide important scientific experiments requiring current leads, e.g. the LHC at Cern and ITER.

For nuclear fusion applications, the applicability of HTS current leads was firstly demonstrated with the 70 kA HTS current lead Demonstrator for ITER [HAA04, HDD05, HFL06]. Following the innovations introduced by this demonstrator, the present ITER current leads as well as the HTS current leads for the stellarator Wendelstein 7-X and the tokamak JT-60SA [FHK09] have been designed and are under construction (ITER, JT-60SA) or have already been built (W7-X). Also the particle accelerator LHC at CERN operates successfully with HTS current leads rated at 13 kA [BMM03, Bal12].

In the next section, the characteristic structure of a HTS current lead is presented. Although the examples mostly refer to the KIT design of the HTS current leads for W7-X and the JT-60SA, the description of the HTS current lead structure has a general validity, also outside the nuclear fusion technology. The discussion about the optimization of a HTS current lead is presented in Chapter III.

2.3 HTS current leads

The general structure of a HTS current lead is presented in Fig. 2.2 and can be summarized as a connection of four main components in series. At the room temperature end of the lead, the room temperature terminal connects the power supply to a resistive conductor, typically made of copper or aluminium, operating in the temperature range between 65/70K - 300 K; this section of the HTS current lead needs to be actively cooled and it is normally referred to as *heat exchanger*. In the temperature range 4.5 K - 65/70 K the electrical current is transported in the HTS module; at the cold end, the low temperature connection provides the current transport to the LTS current distribution line (or bus bar), which are physically connected to the superconducting device.

2.3.1 Room temperature terminal and heat exchanger

The room temperature terminal and the heat exchanger constitute the normal or conventional conducting section of a HTS current lead. They operate over the largest temperature difference, i.e. about 3/4 of the total temperature change. The design of

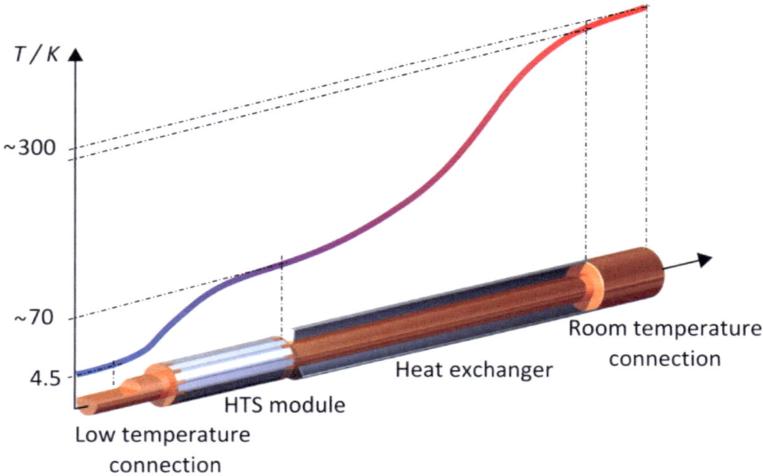

Fig. 2.2: Schematic view of the structure of a HTS current lead and its typical temperature distribution.

these components has to consider two competing contributions, as discussed in section 2.1.1: the Joule heating and the heat conduction. An optimized design should in principle provide a temperature distribution with zero gradient at both the warm end and the interface between the heat exchanger and the HTS module, as shown in Fig. 2.2. The room temperature terminal operates over a relatively small temperature range ($\Delta T \sim 10$ K) and has to be connected to the power grid. It is possible to shape it in such way to have relatively small current densities (in the order of 1 MA/m^2). The balance of the heat conduction and the Joule heating is feasible, but the condition of null gradient at the warm end is realizable only for a specific temperature and current value. The conditions in the external environment may vary, for instance depending on the season; therefore some device for the thermal stabilization of the room temperature terminal end might be required (e.g. heaters). The room temperature terminal of the W7-X HTS current lead prototype is shown in Fig. 2.3.

The heat exchanger operates over a temperature range of about $\Delta T \sim 200$ K. In this case, due to the electrical current to be transported (in the order of tens of kA) as well as to assembling constraints typical of fusion applications, the current density is in the order of 10 MA/m^2. Under these conditions, it is not reasonably possible to design a

Fig. 2.3: Room temperature terminal of the W7-X HTS current lead prototype.

heat exchanger without active cooling and without any heat flux towards the HTS module. The active cooling of the heat exchanger is therefore mandatory.

Basically, two cooling strategies are available: forcing gaseous helium to flow into the heat exchanger; or evaporating liquid nitrogen from a bath at a slightly higher pressure than the atmospheric pressure. In both cases, the inlet of the helium gas and the liquid nitrogen bath are displaced at the cold end of the heat exchanger. It is worth noting that the temperature at the cold end of the heat exchanger must be close to 77 K in case liquid nitrogen is used. This means that the temperature at the transition between heat exchanger and HTS module is sensibly higher than indicated above. Up to today, the forced convection with helium gas has been widely used; nevertheless, the concerns about its availability in the future [BH13], the increasing price as well as the possibility to increase the low temperature limit at the heat exchanger cold side make the use of liquid nitrogen more attractive.

In order to enhance the heat transfer to the coolant, the heat exchanger is typically manufactured with an extended cooling surface. An example is shown in Fig. 2.4.

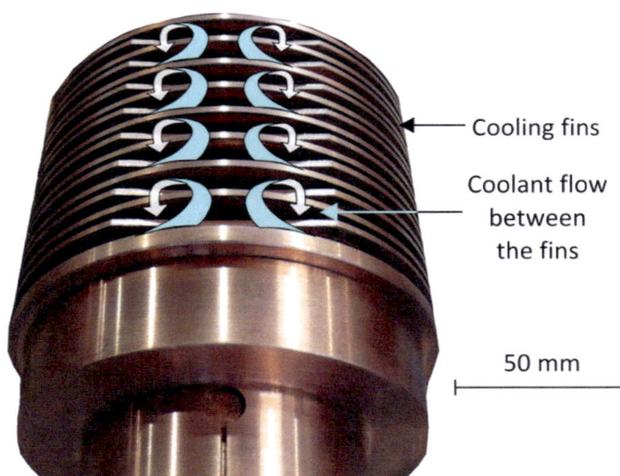

Fig. 2.4: Cooling fins machined on the heat exchanger for the W7-X HTS current lead. The helium is forced to a meander flow between the fins (blue arrows) that enhances the heat transfer from the copper.

2.3.2 HTS module

The HTS module is the main superconducting element of a HTS current lead. Typically this component does not require active cooling. In normal operation, the heat loss is mainly due to the conductive heat flux related to the temperature gradient along the HTS module itself. The heat flux is directed towards the cold end of the HTS module.

The basic structure of an HTS module consists of a hollow stainless steel cylinder on which the high temperature superconductors are housed and fixed by soft soldering. At the two ends of the stainless steel support, two copper caps provide the electrical connection to the other parts of the lead, as shown in Fig. 2.5 a).

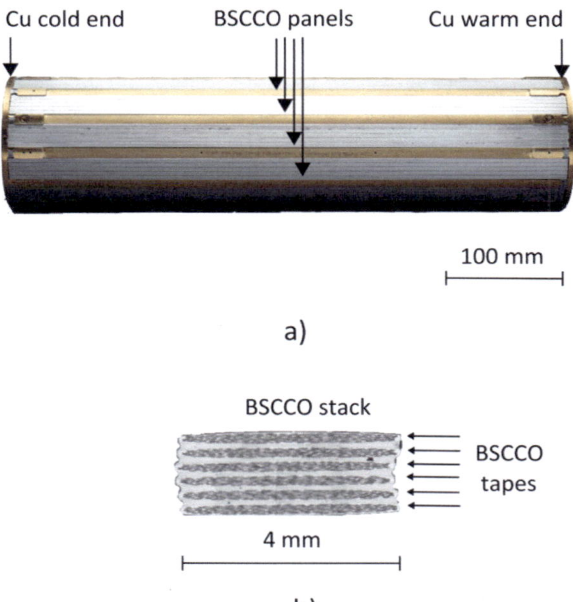

Fig. 2.5: a) HTS module for the W7-X HTS current leads;
b) Cross section of a stack made of six BSCCO-2223 Ag/Au tapes.

Since the current leads usually experience a low magnetic field (in the order of a few tenths of tesla), it is possible to adopt commercial high temperature superconductors: for the present generation of HTS current leads, tapes of BSCCO-2223/AgAu have been soldered together into stacks, and stacks have been jointed into panels to be housed onto the stainless steel cylinder. Figure 2.5 b) summarizes this procedure for the case of W7-X HTS module, where stacks have been assembled by soldering six BSCCO-2223/AgAu tapes and panels are made of five stacks each. In total twelve panels have been used.

2.3.3 Low temperature connection

The last main component of a HTS current lead is the low temperature connection. It connects the cold end of the HTS module to the bus bar. As the HTS module, the low temperature connection is not actively cooled: the temperature is maintained at about 5 K thanks to the bus bar, which is cooled by supercritical helium. An example of a low temperature connection is shown in Fig. 2.6. Although it is normally realized with inserts of LTS material (NbTi, Nb_3Sn), the presence of some normal conductor increases the electrical resistivity. Heat losses are mainly due to the Joule heating, since the temperature gradient along this component is negligible.

2.3.4 Joints and other interconnections

It is worth mentioning that, besides the main components described above, a HTS current lead contains a series of resistive joints and interconnections that connect the different parts of the lead itself. These joints generate thermal losses that need to be precisely evaluated; indeed, proper cooling of all HTS current lead components is required to avoid potentially dangerous heat depositions inside the cryostat.

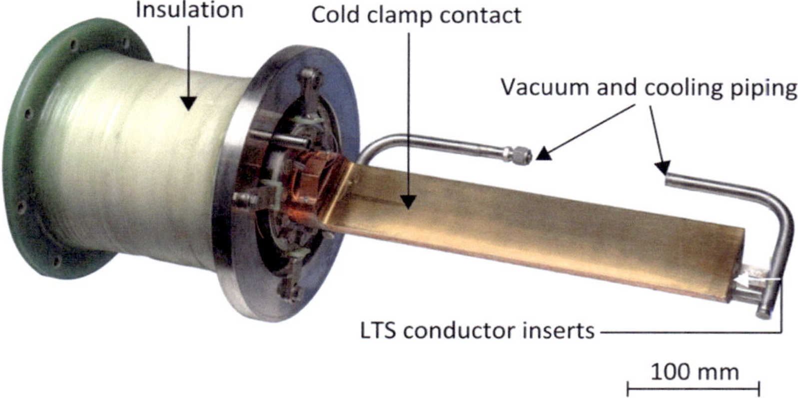

Fig. 2.6: Low temperature connection of the W7-X HTS current leads.

3 Optimization of High Temperature Superconductor current leads

In this section, an overview on the optimization procedure for HTS current leads is presented. The aim is to provide the reader with the essential background for motivating the research discussed in the next Chapters.

3.1 Optimization procedure

Since their very early stage, cryogenic sciences and technologies have largely relied on current leads. The design of current leads is focused on the minimization of the heat load into the cryostat and of the cooling power required to operate the current leads themselves. This procedure has been referred to as *optimization* of a current lead.

Although a HTS current lead requires maximum one third as much cooling power as a conventional current lead [Mum89], the need for optimization is still a central challenge of the design. Indeed, operating a non-optimized HTS current lead would lead to higher, undesired heat loads as well as to higher operation costs (as for conventional current leads); moreover, in case of failure, a sudden and uncontrolled thermal runaway could damage the HTS conductor of the HTS current lead and, possibly, the cryostat equipment and the superconducting device.

3.1.1 Mathematical formulation for the thermal optimization of HTS current leads

The optimization procedure is based on a rigorous mathematical formulation for the cooling problem of a HTS current lead. This procedure has been widely discussed and it is nowadays well accepted. In the following, the fundamentals of this procedure as presented in [Tan06, p.82] are shortly discussed; an outlook in shown in Fig. 3.1.

According to the general overview of a HTS current lead structure presented in Chapter II, the required cooling power to operate the current lead, P_t, consists of the sum of two contributions:

$$P_t = P_{HTS} + P_{HX}, \tag{3.1}$$

where P_{HTS} is the cooling power which has to be provided at the 4.5 K end and P_{HX} is the cooling power for cooling the heat exchanger of the current lead.

In detail, P_{HTS} can be written as:

$$\begin{aligned} P_{HTS} &= \left(\dot{Q}_{HTS,cond} + \dot{Q}_{HTS,loss}\right) \cdot \frac{1}{\eta_{c,1}} \cdot \frac{1}{e_1} \\ &= \left(\frac{A_{HTS}}{L_{HTS}} \cdot \int_{T_{HTS,C}}^{T_{HTS,W}} \lambda(T) \cdot dT + \dot{Q}_{HTS,Loss}\right) \cdot \frac{1}{\eta_{c,1}} \cdot \frac{1}{e_1}, \end{aligned} \tag{3.2}$$

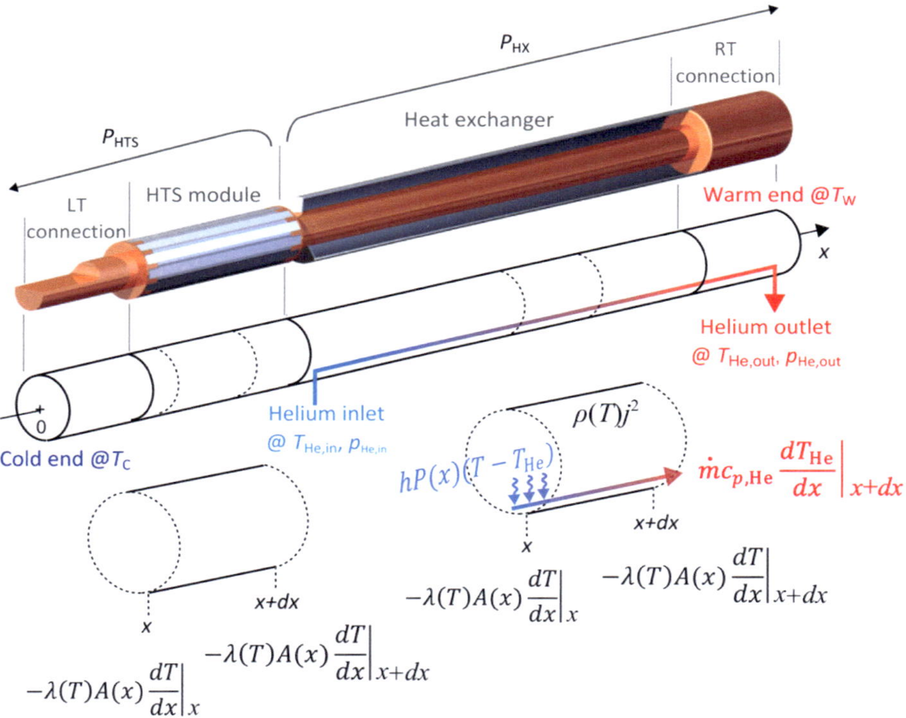

Fig. 3.1 Fundamentals of the mathematical treatment for the cooling of a HTS current lead.

where $\dot{Q}_{\mathrm{HTS,cond}}$ accounts for the heat conducted along the HTS module and $\dot{Q}_{\mathrm{HTS,loss}}$ for the resistive losses at the low temperature end of the current lead. The parameters A_{HTS} and L_{HTS} are the cross section and the length of the HTS module, λ(T) is the equivalent heat conductivity, whereas $\eta_{\mathrm{c},1}$ and e_1 are the Carnot and the refrigerator efficiencies at 4.5 K, respectively.

The formulation for P_{HX}, in case the heat exchanger is cooled with gaseous helium, is:

$$P_{\mathrm{HX}} = \dot{m} \cdot (T_0 \cdot \Delta s_{\mathrm{He}} - \Delta h_{\mathrm{He}}) \cdot \frac{1}{\eta_{\mathrm{c},2}} \cdot \frac{1}{e_2}, \tag{3.3}$$

where $\dot{m}$ is the helium mass flow rate and $\eta_{\mathrm{c},2}$ and e_2 are the Carnot and the refrigerator efficiencies for cooling the helium from room temperature down to the temperature at which it enters the heat exchanger. The terms Δs_{He} and Δh_{He} are the helium entropy and enthalpy differences between the inlet and the outlet of the heat exchanger (the entropy difference is evaluated with respect to the reference temperature T_0, normally $T_0 = 298.15$ K):

$$\Delta s_{\mathrm{He}} = s_{\mathrm{He,out}}\left(p_{\mathrm{He,out}}, T_{\mathrm{He,out}}\right) - s_{\mathrm{He,in}}\left(p_{\mathrm{He,in}}, T_{\mathrm{He,in}}\right), \tag{3.4}$$

$$\Delta h_{\mathrm{He}} = h_{\mathrm{He,out}}\left(p_{\mathrm{He,out}}, T_{\mathrm{He,out}}\right) - h_{\mathrm{He,in}}\left(p_{\mathrm{He,in}}, T_{\mathrm{He,in}}\right). \tag{3.5}$$

Besides the limiting temperatures between which the HTS current lead is operated and the refrigerators' efficiency, the total cooling power P_{t} depends upon the temperature profile along the current lead, upon the coolant mass flow rate $\dot{m}$, upon the temperature change ($T_{\mathrm{He,out}}$ - $T_{\mathrm{He,in}}$) and upon the pressure drop ($p_{\mathrm{He,out}}$ - $p_{\mathrm{He,in}}$) in the coolant. Therefore, for the minimization of the cooling power, the following additional equations are necessary:

$$0 = \lambda(T) \cdot A(x) \cdot \frac{d^2T}{dx^2} + \rho(T) \cdot J^2 \cdot A(x) + h(\dot{m}, T_{\mathrm{He}}, p_{\mathrm{He}}, geom.) \cdot P(x) \cdot (T - T_{\mathrm{He}}), \tag{3.6}$$

$$\dot{m} \cdot c_{p,\mathrm{He}} \cdot \frac{dT_{\mathrm{He}}}{dx} = h(\dot{m}, T_{\mathrm{He}}, p_{\mathrm{He}}, geom.) \cdot P(x) \cdot (T - T_{\mathrm{He}}), \tag{3.7}$$

$$\frac{dp_{\mathrm{He}}}{dx} = \frac{\zeta(\dot{m}, T_{\mathrm{He}}, p_{\mathrm{He}}, geom.)}{dx} \cdot \frac{1}{\rho_{\mathrm{He}}} \cdot \frac{\dot{m}^2}{A_{\mathrm{He}}^2}. \tag{3.8}$$

Equation 3.6 is the one dimensional, time-independent differential equation for the energy balance over the entire current lead. It consists of three contributions: the diffusive microscopic energy transport due to the conduction, the source of thermal energy due to the joule heating and the sink due to the helium cooling, respectively. The latter contribution trivially vanishes for the components that are not actively cooled, e.g. the HTS module. The following quantities are involved in Eq. 3.6: λ(T) and ρ(T) are the equivalent heat conductivity and equivalent electrical resistivity, A(x) and P(x) are the cross section of the lead and the cooling perimeter at a given axial position x, J is the electrical current density and h is the heat transfer coefficient; the heat transfer coefficient h depends upon the mass flow rate $\dot{m}$, the thermodynamic condition of the coolant, i.e. upon its temperature T_{He} and pressure p_{He}, and upon the geometry of the heat exchanger Although the solution of a multidimensional differential equation does not represent a prohibitive task nowadays, it is well established that the 1-D analysis is sufficiently accurate for the investigation of a (HTS) current lead. This is due to the transport phenomena, which mainly occur along the axial direction of the current lead. The assumption of time independence does not lead to any loss of generality either. Even though a HTS current lead can be operated in steady state (J = constant) or periodic pulse mode ($J = f$(time)), it can be optimized only for a specific value of current density J. The steady state analysis is sufficient if the frequency of the pulse is small enough, i.e. much lower than 1 Hz, as it is for the case of HTS current leads for fusion applications.

Equation 3.6 is coupled via the sink term to Eq. 3.7, or to the one dimensional, time-independent differential equation for the energy balance of the coolant. This differential equation consists of two terms: the macroscopic energy transport, or convective term, and the source term. In this equation $c_{\mathrm{p,He}}$ is the helium specific heat capacity.

Equation 3.8 is a rearranged form of the momentum equation for the coolant and relates the flow parameters of the coolant to its pressure loss. The term $\zeta(\dot{m}$, T_{He}, p_{He}, geom.) is the pressure drop coefficient, ρ_{He} is the helium density and A_{He} is the cross section of the cooling channel in which the helium flows.

Equations 3.1 – 3.8 constitute the mathematical formulation for the thermal optimization of HTS current leads. The solution to this problem has already been sought in a variety of different ways, e.g. analytical solutions or functional analysis of a simplified version of Eq. 3.6, but the numerical approach is nowadays the leading choice. Indeed, the numerical solution of such system of equations does not represent a prohibitive task in term of CPU time and computational power. Furthermore, in opposition to analytical solutions, it does not require simplifying assumptions, but only information about the efficiency of the refrigerator(s), the HTS current lead geometry, the material properties and the thermal hydraulics of the coolant.

3.1.2 Thermal-hydraulics of the coolant

The thermal-hydraulics of the coolant is characterized in Eqs. 3.7 and 3.8 by the quantities h and ζ, or the heat transfer and pressure drop coefficient, respectively. These parameters do not depend upon the type of coolant, i.e. they cannot be classified as material properties, but rather on the flow conditions. In turn, the flow conditions depend upon the mass flow rate $\dot{m}$, the thermodynamic state of the coolant (in this case helium, therefore on T_{He} and p_{He}) and on the geometry of the duct in which the coolant flows, i.e. the geometry of the heat exchanger. The mass flow rate $\dot{m}$ and the geometry of the heat exchanger are input parameters to the afore-described optimization model, whereas the thermodynamic state of the coolant is a part of its solution. A numerical code for the optimization of HTS current leads should therefore be provided with generalized functions to derive the heat transfer and the pressure drop coefficients h and ζ on the basis of the computed flow conditions.

The Karlsruhe Institute of Technology, KIT, has adopted a helium cooled heat exchanger with characteristic meander flow geometry, as shown in Fig. 2.3 and, in more detail, in Fig. 4.1, for the HTS current leads of the stellarator W7-X and of the tokamak JT-60SA. Although this kind of heat exchanger has already been used in the HTS current leads for the LHC at CERN and will be mounted in the HTS current leads for ITER as well, an exhaustive knowledge on how the heat transfer and the pressure drop coefficients h and ζ are related to the helium flow conditions is, to the best of the author's knowledge, not yet available.

The thermal-hydraulics of the helium in the meander flow heat exchanger has therefore been extensively investigated by means of a Computational thermal Fluid Dynamics, CtFD, analysis. This analysis technique has been developed and validated against experimental results on short samples of the meander flow heat exchanger [SCF10] and on a full size heat exchanger belonging to the W7-X HTS current leads prototype [RHS11]. Consequently, the influence of the meander flow geometry, the mass flow rate and the helium thermodynamic state have been investigated.

The details about this analysis technique, the methodology and the results are discussed in Chapters IV and V.

3.2 Design tool for the HTS module

Equation 3.1 shows how the cooling power to be provided at 4.5 K depends upon the heat conducted along the HTS module $\dot{Q}_{\mathrm{HTS,cond}}$ and upon the resistive losses $\dot{Q}_{\mathrm{HTS,loss}}$ occurring within it.

The heat conducted along the HTS module is a function of the temperature at the boundaries of the HTS module itself ($T_{\mathrm{HTS,w}} - T_{\mathrm{HTS,c}}$), of the geometry $A_{\mathrm{HTS}}/L_{\mathrm{HTS}}$ and of the material used via $\lambda(\mathrm{T})$. Its contribution can be calculated with Eq. 3.6.

The resistive losses are due to not negligible resistive contributions: the electrical current transport in the HTS module mainly occurs in the HTS conductors; nevertheless transition regions between the HTS module and the heat exchanger or the low temperature connection as well as from normal conductor to superconductor exist. Small resistive losses can therefore not be avoided, thus they shall be properly quantified. This procedure involves the solution of an electrical model for the HTS module, which is not contained in the system of Eqs. 3.1 – 3.8. Indeed, Eq. 3.1 has to be provided with an input for the thermal losses $\dot{Q}_{\mathrm{HTS,loss}}$.

Moreover, the HTS module is the most delicate component of a HTS current lead and the design must ensure safety margins in case of failure operation. A representative example is the occurrence of a Loss Of Flow Accident, LOFA, during which the active cooling of the resistive part of the HTS current lead fails while the current lead is still in operation. Under these conditions, the released heating power is stored in the HTS current lead itself and the temperature increases. The HTS module must be designed to withstand the temperature increase for a prescribed period of time and prevent the HTS conductors from quenching. Indeed, a thermal runaway could seriously occur with potential severe consequences on the device served by the HTS current lead and on the HTS current lead itself.

For this case as well, the optimization model proposed above is not sufficient and has to be integrated with:

- a design tool for a precise steady-state and time dependent thermal-electrical modelling of the HTS module,
- the time dependent version of Eq. 3.6, shown hereunder

$$\frac{\partial T}{\partial t} = \frac{\lambda(T) \cdot A(x) \cdot \frac{\partial^2 T}{\partial x^2} + \rho(T) \cdot J^2 \cdot A(x) - h(\dot{m}, T_{\mathrm{He}}, p_{\mathrm{He}}, geom.) \cdot P(x) \cdot (T - T_{\mathrm{He}})}{\rho(T) \cdot c_{\mathrm{p}}(T)}. \tag{3.9}$$

An exemplarily course of actions for designing a HTS module based on this set of tools would be:

- make a first estimation of the HTS module design and its influence on the performance of the entire HTS current lead by solving Eqs. 3.6-3.9,
- refine and improve the design of the HTS module with the thermal-electrical modelling tool.

Of course, this course of action can be repeated iteratively to find the best solution.

The development of the thermal-electrical modelling tool for the HTS module, its validation and its results will be discussed in Chapter VI.

4 CtFD analysis of the helium flow in the meander flow heat exchanger

This Chapter describes the technique adopted for the Computational thermal-Fluid Dynamic, CtFD, analysis of the helium flow inside the meander flow heat exchanger. Some introductive remarks will provide the reader with the necessary background to understand the rationale behind the choice of this particular computational technique; in the following, the technique as well as the analysis procedure will be described in detail. The content of this Chapter has been published by the author in [RHS13a, RHS13b].

4.1 Meander flow heat exchanger

KIT has chosen a so-called meander flow heat exchanger for the HTS current leads of the stellarator W7-X and the tokamak JT-60SA. This kind of heat exchanger is manufactured from a copper cylinder on which grooves are machined in order to shape fins. Each fin is alternatively cut on the right or on the left external side. The resulting arrangement is a copper bar provided with cut cylindrical fins; the heat exchanger is finally encapsulated in a hollow cylindrical case normally made of stainless steel. The electrical current flows inside the central bar; the helium flows between two neighbour fins, in a so-called layer, in cross motion with respect to the central bar. The transit from a layer to the following one is realized at the external region of the layer. The helium is therefore forced to a fully 3-D meander flow inside this kind of heat exchanger. Exemplarily representations of a section of the meander flow heat exchanger and of the coolant flow are shown in Fig. 4.1 a) and 4.1 b).

The characteristic geometry of this heat exchanger will be referred to as meander flow geometry throughout this work. The meander flow geometry is fully defined by five geometrical quantities:

- the outer diameter, d_o,
- the central bar diameter, d_i,
- the fin distance, t,

- the fin thickness, s, and
- the cut-off distance, c_o.

The relevant geometrical data of the heat exchanger for the W7-X and JT-60SA HTS current leads are given in Tab. 4.1.

Among these quantities the central bar diameter d_i influences the cross section of the conductor; thus it influences the balance between the conducted heat and the Joule heat; the outer diameter d_o, the fin distance t and the fin thickness s influence the extension of the cooling surface of the heat exchanger and the efficiency of the fins. The cut off c_o plays an important role for the pressure drop. All these geometrical quantities influence the shape of the channel in which the helium flows; therefore they influence the heat transfer coefficient h and the pressure drop coefficient ζ, as introduced in Chapter 3.

4.2 CtFD analysis based on the periodic modelling

Introductory considerations on the thermal-hydraulics inside the meander flow heat exchanger

The heat transfer coefficient h and the pressure drop coefficient ζ play a central role in determining the cooling power to be provided at 65 K in order to properly operate the HTS current lead.

Both the heat transfer and the pressure drop coefficients h and ζ depend upon the helium mass flow rate $\dot{m}$, the thermodynamic conditions of the coolant (p_{He} and T_{He}) and the geometry, or, for the meander flow geometry, on d_o, d_i, t, s and c_o. The relations connecting these quantities are normally given in form of correlations of dimensionless numbers.

The helium mass flow rate, the thermodynamic conditions and the geometry are combined into the well-known, dimensionless quantity Reynolds number, Re.

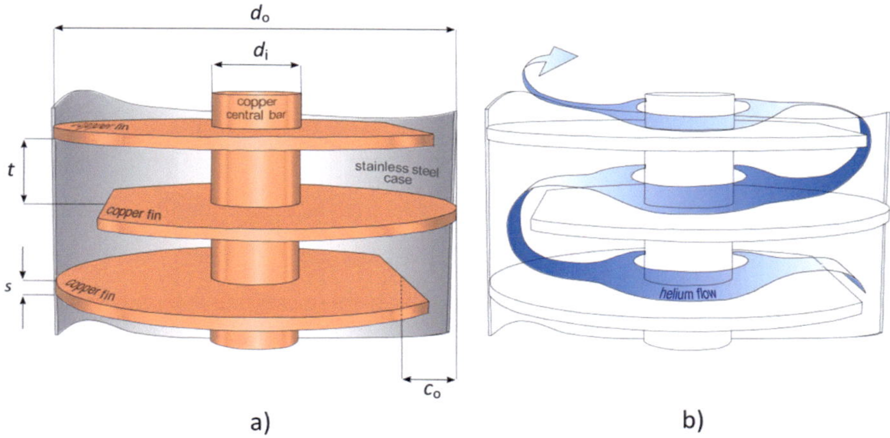

Fig. 4.1: a) Section of a meander flow heat exchanger encapsulated inside a stainless steel case; b) Qualitative representation of the helium flow inside the meander flow geometry.

Tab. 4.1: Data of the meander flow heat exchanger for the HTS current leads of W7-X and JT-60SA. Source [FHK09].

(*) The central bar diameter varies along the heat exchanger's length.

	W7-X, 18.2 kA	**JT-60 SA, 20 kA**	**JT-60 SA, 26 kA**
d_o / mm	120	120	130
d_i / mm	45, 83% of L (*) 55, 17% of L (*)	45, 83% of L (*) 55, 17% of L (*)	50, 81% of L (*) 65, 19% of L (*)
t / mm	3	3	3
s / mm	2	2	2
c_o / mm	7	7	7
L / mm	1207	1207	1077

It is used to characterize the flow itself and it is the most important dimensionless quantity describing the hydraulic analogy for forced convection: flows at the same *Re* number behave similarly, independently of the specific mass flow rate, the thermodynamic conditions or the geometry of the channel. The Reynolds number *Re* is defined as:

$$Re = \frac{\rho_{\mathrm{He}}(p_{\mathrm{He}}, T_{\mathrm{He}}) \cdot |\boldsymbol{v}| \cdot d_{\mathrm{h}}}{\mu_{\mathrm{He}}(p_{\mathrm{He}}, T_{\mathrm{He}})}, \tag{4.1}$$

where ρ_{He} and μ_{He} are the helium density and the molecular viscosity, respectively; both of them are function of the helium thermodynamic conditions, or p_{He} and T_{He}. The quantity $|\boldsymbol{v}|$ is the modulus of the velocity representative of the helium flow inside the channel at p_{He} and T_{He}; d_{h} is the hydraulic diameter of the channel.

In Eq. 4.1, the dependence of *Re* on the helium thermodynamic conditions (via p_{He} and T_{He}) and on the geometry (via d_{h}) can be clearly seen. To highlight the dependence of the Reynolds number *Re* on the helium mass flow rate $\dot{m}$, Eq. 4.1 has to be slightly rearranged. Firstly, the formulation of the mass flow rate is introduced:

$$\dot{m} = \rho_{\mathrm{He}}(p_{\mathrm{He}}, T_{\mathrm{He}}) \cdot \int_{A_{\mathrm{He}}} \boldsymbol{v} \cdot n \, da_{\mathrm{He}}, \tag{4.2}$$

where A_{He} is the cross section of the channel in which the helium is flowing, which is representative of the specific geometry. The term $\boldsymbol{v}$ is the velocity of the helium flowing through A_{He}. Equation 4.2 can be rearranged by introducing a value representative of the velocity magnitude:

$$\int_{A_{\mathrm{He}}} \boldsymbol{v} \cdot n \, da_{\mathrm{He}} = |\boldsymbol{v}| \cdot A_{\mathrm{He}}, \tag{4.3}$$

which leads to

$$\dot{m} = \rho_{\mathrm{He}}(p_{\mathrm{He}}, T_{\mathrm{He}}) \cdot \int_{A_{\mathrm{He}}} \boldsymbol{v} \cdot n \, da_{\mathrm{He}} = \rho_{\mathrm{He}}(p_{\mathrm{He}}, T_{\mathrm{He}}) \cdot |\boldsymbol{v}| \cdot A_{\mathrm{He}}. \tag{4.4}$$

By substituting Eq. 4.4 into Eq. 4.1, the following formulation is obtained:

$$Re = \frac{\dot{m} \cdot d_{\mathrm{h}}}{\mu_{\mathrm{He}}(p_{\mathrm{He}}, T_{\mathrm{He}}) \cdot A_{\mathrm{He}}}. \tag{4.5}$$

This formulation of the Reynolds number *Re* highlights the dependence upon the helium mass flow rate $\dot{m}$, upon the thermodynamic conditions and upon the geometry of the channel in which it is flowing. In more detail, the geometry dependence is given by the ratio $d_{\mathrm{h}}/A_{\mathrm{He}}$. The quantities d_{h} and A_{He} are related to the helium flow field in the channel. At present, the helium flow field in the channel is not known; indeed it is a part of the solution the CtFD analysis is aiming at. Therefore, formulations for the quantities d_{h} and A_{He} consequently the ratio $d_{\mathrm{h}}/A_{\mathrm{He}}$ will be derived in the post-processing phase. To achieve this goal, it is necessary to analyse different meander flow geometrical arrangements, i.e. different 5-tuples of the parameters d_{o}, d_{i}, t, s and c_{o} under different thermodynamic conditions as well as mass flow rates.

Nusselt number, *Nu*

The heat transfer coefficient h is normally arranged in form of the dimensionless quantity called Nusselt number, Nu. The Nusselt number expresses the ratio of the convective to the conductive heat transfer occurring in direction perpendicular to the fluid flow. It is defined as:

$$Nu = \frac{h \cdot d_{\mathrm{h}}}{\lambda_{\mathrm{He}}(p_{\mathrm{He}}, T_{\mathrm{He}})}, \tag{4.6}$$

where λ_{He} is the thermal conductivity of the helium and depends on its thermodynamic state (i.e. on p_{He} and T_{He}). The Nusselt number Nu is generally written in form of a function of the Reynolds number Re and, possibly, of other dimensionless quantities[4].

Pressure drop coefficient, ζ

The pressure drop coefficient is a dimensionless quantity which relates the pressure drop occurring over a specific section of the channel to the flow velocity square $|\boldsymbol{v}|^2$ and the fluid density ρ_{He}. The formulation derives from Bernoulli equation with no potential energy contribution and can be written as:

$$\zeta = \frac{\Delta p_{\mathrm{He}}}{2 \cdot \rho_{\mathrm{He}}(p_{\mathrm{He}}, T_{\mathrm{He}}) \cdot |\boldsymbol{v}|^2}. \tag{4.7}$$

By introducing Eq. 4.4 in Eq. 4.7:

[4] Typically the Nusselt number is correlated also to the Prandtl number, Pr. The Prandtl number is the ratio between the viscous diffusivity and the heat diffusivity. In this work it is not considered since its total variation is less than 4% over the ranges of T_{He} and p_{He} in Tab. 4.2; therefore its contribution in defining more precisely Nu is negligible.

$$\zeta = \frac{\Delta p_{He} \cdot \rho_{He}(p_{He}, T_{He}) \cdot A_{He}^2}{2 \cdot \dot{m}^2}. \tag{4.8}$$

Similarly to the Nusselt number Nu, the pressure drop coefficient ζ can be written in form of a function of the Reynolds number Re and, possibly, of other dimensionless quantities.

4.2.1 Scope of the CtFD analysis

The CtFD analysis of the meander flow heat exchanger aims at highlighting the relations between the helium flow conditions, expressed by the Reynolds number Re, and the heat transfer coefficient h (via the Nusselt number Nu) and between Re and the pressure drop coefficient ζ. The goal is to define functions as $Nu = f(Re, x_i)$ with i=1,2…n and $\zeta = g(Re, x_j)$ with j=1,2…n discussed above. The terms x_i and x_j are dimensionless quantities derived from ratios of meander flow geometry parameters. The procedure is as follow: firstly it is necessary to define the ranges in which these functions have to be found, or the ranges for the mass flow rate, the thermodynamic conditions and the meander flow geometrical parameters introduced in § 4.1. Then a technique for the CtFD analysis must be chosen and applied.

4.2.2 Ranges for the CtFD analysis

For the sake of generality, the ranges for the CtFD analysis have been defined in order to cover the parameters of the HTS current leads mounting a meander flow heat exchanger and being presently designed or built: i.e. the HTS current leads for W7-X, for JT - 60SA and for ITER. The ranges of interest are summarized in Tab. 4.2.

According to Tab. 4.2, a set of twenty-five meander flow geometries was considered. These geometries were organized in four macro-groups. Each macro-group aims at highlighting the effect of a specific meander flow geometry parameter on the thermal-fluid dynamics of the helium. The first macro-group focuses on the effect of the central bar diameter d_i for different values of the outer diameter d_o, the second focuses on the fins distance t, the third focuses on the fin thickness s, whereas the fourth focuses on the width of the cut-off c_o.

All twenty-five meander flow geometries, the macro-group they belong to as well as the corresponding meander flow geometry parameters are summarized in Tab. 4.3.

The handling of the parameters listed in Tab. 4.2 and Tab. 4.3 for the purpose of the CtFD analysis will be discussed in details later on in this Chapter.

Tab. 4.2: Range of interest for the CtFD analysis.

Meander flow geometry parameter	
Outer fin diameter, d_o / mm	100 - 200
Central bar diameter, d_i / mm	35 - 120
Fin distance, t / mm	2.7 - 8
Fin thickness, s / mm	0.3 - 6
Cut-off width, c_o / mm	2.7 - 20
Thermodynamic / helium flow properties	
Temperature, T_{He} / K	50 - 300
Pressure, p_{He} / MPa	0.2 – 0.5
mass flow rate $\dot{m}$ / g/s	0.2 - 5

Tab. 4.3: List of the twenty-five meander flow geometries on which the CtFD analysis was performed.

Macro-group #	Meander flow geom. #	d_o / mm	d_i / mm	t / mm	s / mm	c_o / mm
I	1	120	45	3	2	7
	2	120	55	3	2	7
	3	120	65	3	2	7
	4	140	45	3	2	7
	5	140	65	3	2	7
	6	160	65	3	2	7
	7	160	80	3	2	7
	8	160	100	3	2	7
	9	160	120	3	2	7
	10	180	90	3	2	20
	11	180	100	3	2	20
	12	200	120	3	2	20
II	13	140	45	5	2	7
	14	140	45	8	2	7
	15	180	90	5	2	20
	16	180	90	8	2	20
III	17	140	45	3	0.3	7
	18	140	45	3	4	7
	19	140	45	3	6	7
IV	20	120	55	3	2	4
	21	120	55	3	2	10
	22	140	45	5	2	4
	23	140	45	5	2	10
	24	100	40	2	2.7	2.7
	25	160	90	3	2	12

4.2.3 Development of the periodic modelling for the CtFD analysis

The CtFD periodic modelling

The choice of the computational technique for the analysis of the helium thermal-fluid dynamics inside the meander flow heat exchanger is not unique. The first and more straightforward method is the modelling of an entire heat exchanger. This approach was adopted at KIT for simulating the meander flow geometry mock-ups; it was also extended to the heat exchanger of the W7-X HTS current lead prototype. More recently, it has also been used at CERN for simulating the meander flow heat exchanger mounted in the HTS current leads for the ITER TF coil [SBB13]. Nevertheless, this approach presents some non-negligible drawbacks, despite its immediacy. Looking at Fig 4.2, it is clear that in the meander flow geometry the most relevant phenomena for the helium thermal-fluid dynamics occur on a length scale of the same order of magnitude of t, at most. The fin distance is normally in the order of a few millimetres, whereas the total length of a normal heat exchanger is in the order of 1 m. The length scales for modelling an entire heat exchanger still catching the relevant physics span over three orders of magnitude, at least. From a computational point of view, this leads to a fine discretization of the computational domain on a local scale, in comparison to the global dimension of the problem; for a specific computational power, the CPU time becomes therefore longer.

Further relevant considerations are listed below:

- the meander flow geometry as well as the channel in which the helium is flowing have a periodic structure, as shown on the left side of Fig. 4.3,
- the results presented in [HCB10] have shown that the helium flow is fully developed about ten layers after the inlet (and remains fully developed till about 10 layers before the outlet),
- the experimental results as well as the numerical simulations have shown that over a single period of the meander flow geometry the temperature and the pressure variations in all directions is limited to a few kelvin and pascal, respectively,

- since the temperature as well as the pressure variation is limited, the helium thermodynamic properties can be considered as constant over a period of the meander flow geometry.

On the basis of these considerations, an alternative computational approach has been developed based on the work done for the analysis of the transverse transport phenomena in the Cable-In-Conduit-Conductor for the superconducting magnet system of ITER [ZGM06]. Instead of solving the thermal-problem for an entire meander flow heat exchanger, a single period of the heat exchanger has been first isolated as shown in the central part of Fig. 4.3. According to the conditions under which a meander flow heat exchanger normally operates, it has been possible to limit the thermal-fluid dynamic analysis to a single period of the helium channel, as shown in Fig. 4.3 on the right side. This computational domain consists only of the fluid, or helium part. It retains all the geometrical features of the meander geometry from which it has been derived; furthermore, since the helium thermodynamic properties can be considered constant and the flow fully developed, also the helium flow features must be same as if the period was actually inside a heat exchanger (provided it is far enough from the inlet and the outlet). This kind of computational approach will be referred to as *CtFD analysis with the periodic modelling* throughout this work.

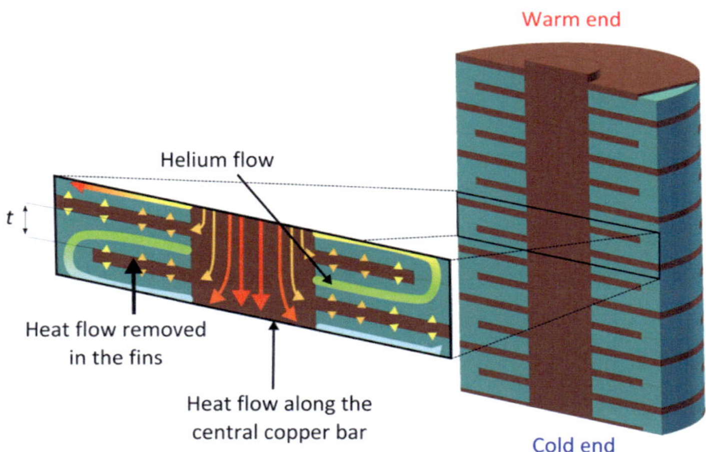

Fig. 4.2: Section of a meander flow heat exchanger. In the blow-up, the curvy arrow indicates the helium flow, whereas the red to yellow faded arrows qualitatively indicate the heat flow from the central copper bar to the fins: the relevant phenomena occur on a t-length scale.

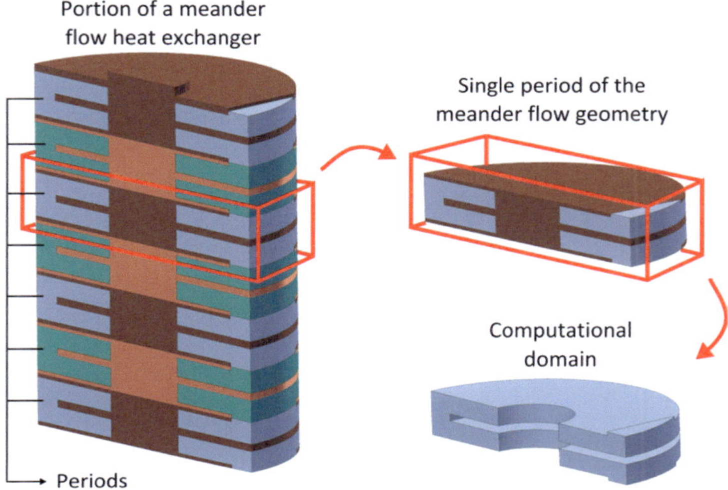

Fig. 4.3: Progressive definition of the computational domain: the periodicity of both the copper heat exchanger as well as of the helium channel has been highlighted on the right side. A period is isolated (in the middle) and from that, the computational domain is reduced to the helium channel, shown on the right side.

Boundary conditions for the CtFD analysis with periodic modelling

The helium inlet and the outlet in the computational domain of the CtFD periodic modelling are modelled with a couple of partially cyclic boundary conditions, as shown in Fig. 4.4 a). This kind of boundary condition requires imposing either the mass flow rate $\dot{m}$ or the pressure drop Δp_{He}, together with the inlet temperature of the helium. The model computes then either the pressure drop, if the mass flow rate has been imposed, or the mass flow rate, if the pressure drop has been imposed. The variation of the helium temperature occurring between the inlet and the outlet can also be computed.

The computational domain is halved by taking advantage of the symmetry of the meander flow geometry, as it can be seen in Fig. 4.1, 4.2 and 4.3. The application of a symmetry plane boundary condition is therefore required, as shown in Fig. 4.4 b).

Since the periodic model does not include any copper part of the heat exchanger, so-called wall boundary conditions are applied at the former solid-fluid interfaces. The set of wall boundary conditions is presented in Fig 4.4 c) - g). The no-slip condition has

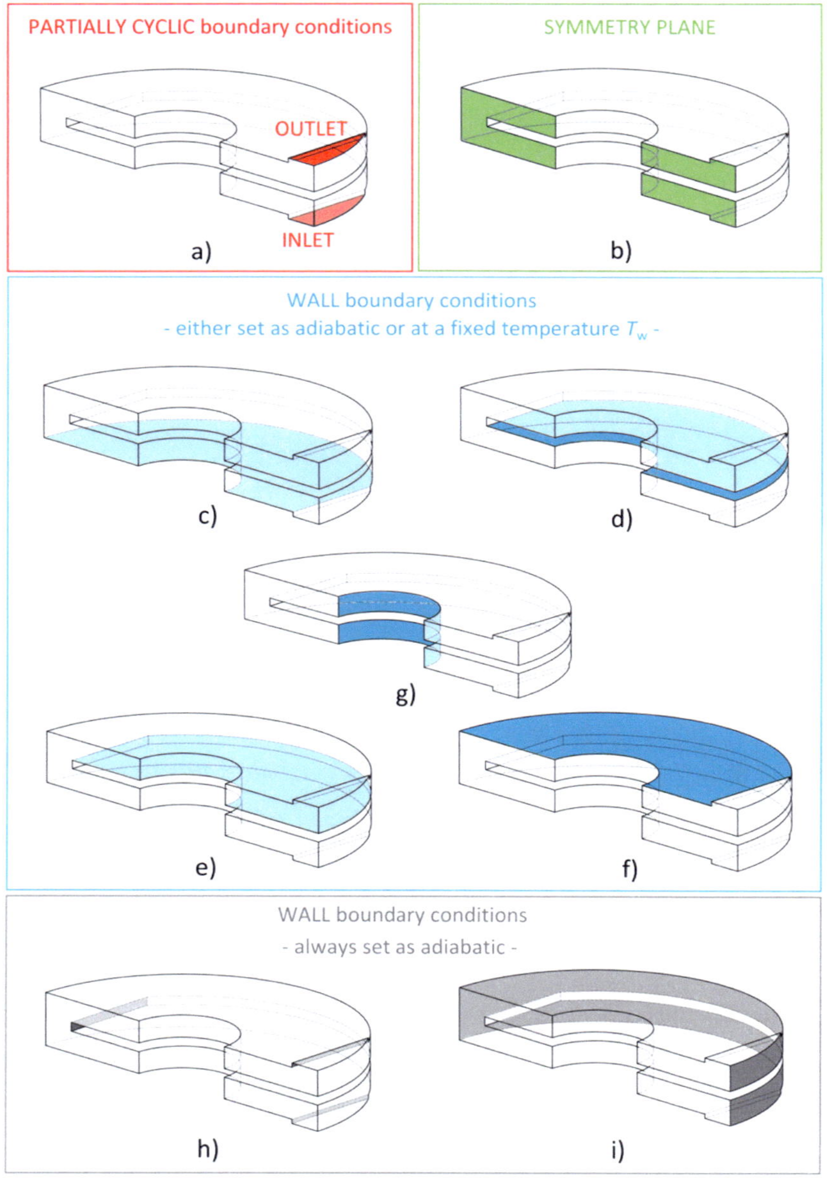

Fig. 4.4: a) Partially cyclic inlet and outlet;
b) Symmetry plane;
c) – g) Wall-type boundary conditions (adiabatic or at a fixed temperature);
h) – i) Wall-type boundary conditions (always adiabatic).

been imposed onto all walls; it is possible to treat them as adiabatic walls or to impose a fixed temperature.

The external cylindrical surface as well as the lateral surface of the fins in the cut-off region is also treated as a wall, see Fig. 4.4 h) and i). Also on these walls the no-slip condition has been imposed; moreover, they are always defined as adiabatic. The assumption of an adiabatic external cylindrical surface holds since in a real HTS current lead the meander flow heat exchanger is encapsulated in a thermally insulated tube. Regarding the assumption of an adiabatic lateral surface in the cut-off region, it has been verified that it does not lead to any major change in the computed results.

4.2.4 Strategy for the CtFD analysis with the periodic modelling

The thermal-fluid dynamic analysis of the helium flow inside the meander flow geometry has to cover several meander flow geometries (see Tab. 4.3) under different helium thermodynamic conditions as well as mass flow rates.

In Tab. 4.3, twenty-five different meander flow geometries have been investigated, which cover the range of meander flow geometrical parameters relevant for this work. For each of these geometries, the computational domain for the periodic modelling has been created. Before going through the physical models solved within the periodic modelling, one has to define under which conditions in terms of helium thermodynamic properties and mass flow rates the models will be run. As stated above, the helium flow conditions are characterized by the Reynolds number *Re*; this latter depends on the mass flow rate, the thermodynamic conditions (via p_{He} and T_{He}) and on the geometry, or on $d_{\mathrm{h}}/A_{\mathrm{He}}$. The terms d_{h} and A_{He} are not known a priori, though the expression for the Reynolds number *Re* can be arranged in form of the dimensional, mass flow rate and thermodynamic properties dependent quantity *Re**.

The formulation for *Re** is:

$$Re^{*} = Re \cdot \frac{A_{\mathrm{He}}}{d_{\mathrm{h}}} = \frac{\dot{m}}{\mu_{\mathrm{He}}(p_{\mathrm{He}}, T_{\mathrm{He}})}. \tag{4.9}$$

The parameter *Re** has the dimension of a length. It is a quantity that allows a more comfortable handling of the mass flow rate and the thermodynamic properties. Of course, it is not possible to compare the helium flow at the same *Re**, but in meander flow geometries with different d_h/A_{He}. On the other hand, for the helium flow inside a specific meander flow geometry, *Re** provides information on the flow analogous to *Re*, but geometry-dependent.

Looking at Tab. 4.2 and considering that the helium molecular viscosity increases with the temperature, whereas its dependence on the pressure is quite weak, it is possible to identify a lower and an upper limit for *Re** within which the possible $\dot{m}/\mu_{He}$ conditions described in Tab. 4.3 are contained. These limits are:

- Re^*_{min} = 10 m, for $\dot{m}$ = 0.2 g/s, T = 300 K and p = 0.5 MPa,
- Re^*_{max} = 800 m, for $\dot{m}$ = 5 g/s, T = 50 K and p = 0.2 MPa.

The helium flow inside each meander flow geometry listed in Tab. 4.3 has been analysed at several *Re** between 10 and 800. In the first place, a solution for each meander flow geometry has been computed at thirteen *Re** positions within the interval 10 - 800. In case a higher resolution over a certain *Re** sub-range was needed, the analysis has been extended to additional *Re** values.

In order to make the analysis as robust as possible, the following approach has been used for the choice of the *Re** positions: first a discrete distribution of *Re** values, called $Re^*_{distribution}$, was given (the twenty-five set of $Re^*_{distribution}$ values used, one for each meander flow geometry listed in Tab. 4.3, differ from each other). Then, for each $Re^*_{distribution}$ value, a random selection of a value for the mass flow rate $\dot{m}_{random}$, one for a temperature, T_{random}, and one for a pressure, p_{random}, within the ranges defined in Tab. 4.2, has been repeated until the ratio of the mass flow rate to the helium molecular viscosity at T_{random} and p_{random}, or $Re^*_{calculated}$, differed from the given $Re^*_{distribution}$ value for less than a defined tolerance (in the order of 1e-01). An example of the results given by this procedure is shown in Tab. 4.4. In the following, the values $Re^*_{calculated}$ will be simply referred to as *Re**.

At this point, two kinds of simulations are performed on all meander flow geometries, at each *Re**:

- a steady-state hydraulic analysis,
- a steady-state thermal-hydraulic analysis.

The aim of the steady-state hydraulic analysis is to compute the helium velocity flow field and to evaluate the pressure drop Δp_{He} occurring between the inlet and the outlet of the periodic model, which is defined as:

$$\Delta p_{\mathrm{He}} = \left| p_{\mathrm{He,out}} - p_{\mathrm{He,in}} \right|. \tag{4.10}$$

In the hydraulic analysis the equations for the mass and momentum conservation are solved. A prescribed helium mass flow rate is imposed between the inlet and outlet, whereas at the inlet the temperature $T_{\mathrm{He,in}}$ and pressure $p_{\mathrm{He,in}}$ are provided. These quantities are set equal to the 3-plet $\dot{m}_{\mathrm{random}}$, T_{random} and p_{random} from which Re^* was calculated. All the wall boundary conditions are treated as adiabatic.

The aim of the steady state thermal-hydraulic analysis is to evaluate the temperature difference ΔT_{He} between the inlet and the outlet of the periodic model, which is defined as:

$$\Delta T_{\mathrm{He}} = T_{\mathrm{He,out}} - T_{\mathrm{He,in}}. \tag{4.11}$$

The thermal-hydraulic analysis is performed in series to the hydraulic analysis and consists in solving the energy conservation equation for the helium. The helium velocity flow field previously computed with the hydraulic analysis is used as an input. The helium inlet conditions are the same as for the hydraulic analysis. On the wall boundary conditions in Fig 4.4 c) - g) a constant temperature T_{w} is imposed, such that $T_{\mathrm{w}} > T_{\mathrm{He,in}}$.

From the computed temperature difference ΔT_{He}, the heat transfer coefficient h can be then derived by performing the heat balance over the computational domain, or:

$$\dot{m} \cdot c_{\mathrm{p,He}} \cdot \Delta T_{\mathrm{He}} = h \cdot A_{\mathrm{HX}} \cdot \left(T_{\mathrm{w}} - T_{\mathrm{b,He}}\right), \tag{4.12}$$

$$h = \frac{\dot{m} \cdot c_{\mathrm{p,He}} \cdot \Delta T_{\mathrm{He}}}{A_{\mathrm{HX}} \cdot \left(T_{\mathrm{w}} - T_{\mathrm{b,He}}\right)}, \tag{4.13}$$

where A_{HX} is the area of the heated surfaces, or the area of the surfaces on which the temperature T_{w} has been imposed; $T_{\mathrm{b,He}}$ is the helium flow bulk temperature, which is the mean temperature of the helium averaged on the flow field inside in the computational domain. The general formulation of the bulk temperature $T_{\mathrm{b,He}}$ can be written as:

$$T_{\mathrm{b,He}} = \frac{\int_{\Sigma} T \cdot \boldsymbol{v} \bullet n \, d\sigma}{\int_{\Sigma} \boldsymbol{v} \bullet n \, d\sigma}. \tag{4.14}$$

How the bulk temperature has been actually calculated in the post-processing phase of the CtFD analysis will be discussed in the next Chapter.

Tab. 4.4: Exemplarily set of Re^* values for the CtFD analysis with the periodic modelling. The random selection of a mass flow rate and thermodynamic conditions has been repeated until $Re^*_{calculated} \sim Re^*_{distribution}$. The CtFD analysis is performed at each Re^* value.

***Re**distribution / m**	13	20	23	26
$\dot{m}_{random}$ / g/s	0.250	0.250	0.290	0.310
T_{random} / K	286.100	149.100	149.100	140.100
p_{random} / MPa	0.405	0.405	0.405	0.203
μ_{He} / (g/(s·m)·1e-02)	1.925	1.251	1.251	1.200
***Re**calculated / m**	12.99	19.99	23.19	25.83
ρ_{He} / kg/m3	0.682	1.309	1.309	0.696
$c_{p,He}$ / J/(kg·K)	5193	5194	5194	5195
λ_{He} / W/(m·K)	0.151	0.096	0.096	0.092

***Re**distribution / m**	51	75	100	130
$\dot{m}_{random}$ / g/s	0.400	1.300	1.900	1.600
T_{random} / K	71.000	245.000	281.100	146.100
p_{random} / MPa	0.405	0.304	0.203	0.203
μ_{He} / (g/(s·m)·1e-02)	7.875	1.733	1.901	1.233
***Re**calculated / m**	50.79	75.00	99.96	129.76
ρ_{He} / kg/m3	2.746	0.597	0.347	0.568
$c_{p,He}$ / J/(kg·K)	5206	5193	5193	5194
λ_{He} / W/(m·K)	0.059	0.135	0.149	0.095

***Re**distribution / m**	185	241	351	483
$\dot{m}_{random}$ / g/s	3.500	3.600	2.900	3.200
T_{random} / K	279.100	197.100	77.070	54.040
p_{random} / MPa	0.304	0.405	0.405	0.203
μ_{He} / (g/(s·m)·1e-02)	1.892	1.500	8.275	6.632
***Re**calculated / m**	184.99	240.00	350.45	482.51
ρ_{He} / kg/m3	0.524	0.990	2.532	1.805
$c_{p,He}$ / J/(kg·K)	5193	5193	5203	5218
λ_{He} / W/(m·K)	185	241	351	483

Important remarks on the strategy

As stated above, the helium material properties are reasonably set constant. However, a density variation always occurs in a real heat exchanger if the temperature or the pressure or both changes.

In order to limit the influence of the assumption of constant properties as much as possible, the following constraint has been applied: the maximum difference between the density values used in the computations and those theoretically derived after the pressure had decreased and/or the temperature had increased was limited at 5%. The pressure drop is generally too small to lead to a density variation of 5% (regardless of the pressure at the inlet of the computational domain); however, this is not always the case for the temperature increment. For this reason, attention has to be paid in setting the values of the wall temperature T_w to apply during the thermal-fluid dynamic analysis. For low helium inlet temperatures, which means from $T_{He,in}$ = 50 K up to $T_{He,in}$ = 100 K, the maximum temperature increment to maintain the density variation below 5% is limited to 2-3 K. On the other hand, according to the range of mass flow rates considered here, low values of the Reynolds number *Re* occur at higher temperature, where the sensitivity of the density to the temperature increment decreases consistently. In these cases, the wall temperature T_w has been limited to be at most 6 K higher than $T_{He,in}$.

Equation 4.12 implicitly assumes that the heat release due to dissipation within the computational domain is negligible. This hypothesis is generally verified. However, if the meander flow geometry under investigation has a large pressure drop (e.g. if the cut-off is small compared to the outer diameter of the plate), some effects of the dissipation on the heat transfer coefficient *h* can be seen if the wall temperature T_w is less than 1-1.5 K higher than $T_{He,in}$; nevertheless, it has been verified that for slightly higher values of (T_w - $T_{He,in}$) the value of *h* stabilizes and becomes independent of (T_w - $T_{He,in}$). Furthermore, the effect of the dissipation is much weaker at lower temperature; hence it is always possible to perform the thermal analysis and to satisfy the constraint on the helium density.

4.2.5 Flow regime and physical models

Up to now, the general procedure for the analysis of the helium thermal-fluid dynamics with the periodic modelling has been presented; no attention has been devoted to the physical models that are actually solved with the numerical analysis. The models for the investigation of the thermal-fluid dynamic depend on the flow regime; therefore it is necessary to know in advance if the helium flow is, at a given *Re**, laminar or turbulent.

Unfortunately, the helium flow regime is not known a priori. Moreover, to the best of the author's knowledge, no experimental evidence has been presented on this topic yet. According to some speculative considerations, one can argue that a laminar flow regime cannot exist in such complicated geometries. Indeed, the sharp changes in the direction of the helium flow as well as the variable cross section of the channel might prevent the existence of any laminar flow. A thermal-fluid dynamics modelling based on the turbulent analysis seems to be, at least in principle, a more suitable approach. On the other hand, at lower *Re** values the turbulent modelling could be inadequate: the helium flow could present a predominant viscous behaviour, typical of the laminar regime. For the sake of completeness and exhaustiveness, the following strategy has been therefore adopted: the hydraulic analysis based on the turbulence model has been run for each meander flow geometry in Tab. 4.2 from the highest towards the lower *Re**. By checking the magnitude of the computed characteristic turbulent flow quantities, it has been possible to identify a lower limit in terms of *Re** for the turbulent flow. At the same time, the hydraulic analysis based on the laminar modelling has been run for each meander flow geometry in Tab. 4.2 from the lowest towards the higher *Re**. The post-processing analysis has then highlighted if such modelling approach was suitable or not for the particular *Re** value and meander flow geometry. Once the hydraulic analysis was completed, the thermal-hydraulic analysis was started based on the same models as the hydraulic analysis.

In the following the physical models solved with the CtFD analysis are presented. Both the model for the laminar flow as well as that for the turbulent flow is based on sets of conservation equations for the mass (also known as continuity equation), the momentum and the energy. The turbulent model contains as many additional conservation equations as the number of turbulent quantities.

Physical model for the laminar flow

For the modelling of the helium thermal-fluid dynamics in laminar flow, the steady state, incompressible Navier-Stokes equations are needed. They derive from the general Navier-Stokes equations, which consist of the continuity equation, the momentum conservation and the energy conservation equation. The Eulerian differential form of the Navier-Stokes equations and the assumptions leading to their steady state, incompressible formulation are given below.

Continuity equation

$$\frac{\partial \rho_{\mathrm{He}}}{\partial t} + \nabla \cdot (\rho_{\mathrm{He}} \cdot \boldsymbol{v}) = 0, \tag{4.15}$$

Momentum conservation equation

$$\rho_{\mathrm{He}} \cdot \left(\frac{\partial \boldsymbol{v}}{\partial t} + \boldsymbol{v} \bullet \nabla \boldsymbol{v}\right) = -\nabla p_{\mathrm{He}} + \mu_{\mathrm{He}} \cdot \nabla \bullet \boldsymbol{T} + \boldsymbol{F}, \tag{4.16}$$

where $\boldsymbol{T}$ is the stress tensor and $\boldsymbol{F}$ the body forces (e.g. the gravity) acting on the fluid.

Energy conservation (temperature formulation)

$$\begin{aligned} \rho_{\mathrm{He}} \cdot c_{\mathrm{p,He}} \cdot \left(\frac{\partial T_{\mathrm{He}}}{\partial t} + \boldsymbol{v} \bullet \nabla T_{\mathrm{He}}\right) &= \\ = \nabla \bullet (\lambda_{\mathrm{He}} \cdot \nabla T_{\mathrm{He}}) + q''' + \beta_{\mathrm{He}} \cdot T_{\mathrm{He}} \cdot \left(\frac{\partial p_{\mathrm{He}}}{\partial t} + \boldsymbol{v} \bullet \nabla p_{\mathrm{He}}\right) &+ \mu_{\mathrm{He}} \cdot \boldsymbol{\Phi}, \end{aligned} \tag{4.17}$$

where q''' is the heat power source density (due, for instance, to chemical or nuclear reactions), β_{He} is the coefficient of thermal expansion of the helium and $\boldsymbol{\Phi}$ is the viscous dissipation function.

The steady state analysis nullifies all the time derivatives appearing in the equations above. Under the conditions shown in Tab. 4.2, the assumption of incompressible flow holds because the Mach number is below 0.3. Also the assumption of constant helium thermodynamic properties does not lead to any major simplification, since the temperature and pressure changes occurring over the computational domain is small enough. Furthermore, considering that the $\boldsymbol{F}$ term in Eq. 4.16 consists only of the gravitational contribution, that q''' is zero because no heat power source is present and that $\boldsymbol{\Phi}$ is negligible, the Eqs. 4.15 – 4.17 reduce to:

$$\nabla \bullet \boldsymbol{v} = 0, \tag{4.18}$$

$$(\boldsymbol{v} \bullet \nabla \boldsymbol{v}) = -\frac{\nabla p_{\mathrm{He}}}{\rho_{\mathrm{He}}} + \upsilon_{\mathrm{He}} \cdot \nabla^2 \boldsymbol{v} + \frac{\boldsymbol{F}}{\rho_{\mathrm{He}}}, \tag{4.19}$$

$$\rho_{\mathrm{He}} \cdot c_{\mathrm{p,He}} \cdot (\boldsymbol{v} \bullet \nabla T_{\mathrm{He}}) = \nabla \bullet (\lambda_{\mathrm{He}} \cdot \nabla T_{\mathrm{He}}), \tag{4.20}$$

where ν_{He} is the dynamic viscosity of the helium.

Equations 4.18 – 4.20 constitute the physical model for the analysis of the laminar helium flow inside the meander flow geometry. The first two equations (4.18 and 4.19) are solved in the hydraulic analysis. This provides the pressure drop occurring across the computational domain as well as the helium flow field. The latter one is used as an input for the solution of Eq. 4.20, which solves the temperature transport problem and provides the temperature difference occurring across the computational domain when $T_{\mathrm{w}} > T_{\mathrm{He,in}}$ is imposed to the walls.

Physical model for the turbulent flow

As physical model for the turbulent flow, the eddy viscosity two-equation k–ω SST model [Men94] has been adopted. In this section the most relevant equations as well as characteristics of this model will be presented. A more detailed explanation about the rationale behind the choice of this model is given in the last section of this Chapter.

A part of the equations constituting the turbulence model used here are equivalent to the Navier-Stokes equations, i.e. they describe the continuity, the momentum conservation and the energy conservation within the fluid flow. Nevertheless, for a better description of turbulent flows, additional hypothesis are needed. Reynolds

[Wil98] proposed to consider the velocity field $\boldsymbol{v}$ as each of its components was made of two parts:

$$\boldsymbol{v} = \boldsymbol{V} + \boldsymbol{v}'. \tag{4.21}$$

The term $\boldsymbol{V}$ is a vector whose components are the time averaged components of the flow field. The term $\boldsymbol{v}'$ is a vector whose components indicate the fluctuations of the velocity field in each direction.

By substituting Eq. 4.21 in Eqs. 4.15 - 4.16 and performing the time average, the following continuity and momentum equations are obtained (already in steady state formulation):

$$\nabla \bullet \boldsymbol{V} = 0, \tag{4.22}$$

$$(\boldsymbol{V} \bullet \nabla \boldsymbol{V}) = -\frac{\nabla p_{\mathrm{He}}}{\rho_{\mathrm{He}}} + \frac{1}{\rho_{\mathrm{He}}} \cdot \nabla \bullet (\mu_{\mathrm{He}} \cdot \nabla \boldsymbol{t} - \rho_{\mathrm{He}} \cdot \boldsymbol{\tau}) + \frac{\boldsymbol{F}}{\rho_{\mathrm{He}}}. \tag{4.23}$$

These equations are generally referred to as the Reynolds-averaged Navier-Stokes equations, RANS [Wil98].

By stating that $\boldsymbol{T} = \nabla \boldsymbol{t}$, it is possible to recognize that the only difference in the momentum equation with respect to Eq. 4.16 is the term $-\rho_{\mathrm{He}} \bullet \boldsymbol{\tau}$. The quantity $-\boldsymbol{\tau}$ comes out from the time averaging process and it is related to the averaging of the oscillating components of the velocity $\boldsymbol{v}$. It is called specific Reynolds stress tensor and is the time-averaged rate of momentum transfer due to the turbulence.

If one assumes that $\nabla \boldsymbol{t}$ and $\boldsymbol{\tau}$ have the same orientation, than the term $\boldsymbol{\tau}$ can be thought as an additional contribution to the viscous dissipation. In this sense, the total viscous dissipation consists of the sum of the material viscosity (in this case μ_{He} if it refers to the molecular viscosity or ν_{He} if it refers to the kinematic one) and the turbulent or eddy viscosity, which is indicated as μ_{t} in case it refers to the molecular eddy viscosity or with ν_{t} in case it refers to the kinematic eddy viscosity.

Several approaches have been proposed to calculate the kinematic eddy viscosity ν_{t} [Wil98]. Most of them put in relation the eddy viscosity to a quantity called turbulent kinetic energy, κ. Nevertheless, a model based on Eq. 4.22, Eq. 4.23 and an equation for κ would not be self-consistent. Indeed, according to dimensions of the turbulent phenomenology (length, time and velocity), an equation for a further quantity is still missing [Wil98]. Historically, models based on an equation for the turbulent kinetic energy and one equation for either the turbulent dissipation ε (κ-ε model) or for the dissipation rate ω (κ-ω model) have been developed. These models have been largely adopted and validated against several types of flow. They are referred to as two-equation models, where the equations for κ and ε or ω are in the form of conservation equations.

The performance of the κ-ε and the κ-ω model are not equivalent; indeed, it is nowadays accepted and verified that the κ-ε model is more suitable to simulate the wake region or a turbulent boundary layer, whereas the κ-ω offers more reliable predictions of the flow region close to the walls. To take advantage of both these features, Menter proposed a novel turbulence model [Men94] based on the mutually exclusive coupling of the κ-ε and the κ-ω model. To reduce the number of variables, the turbulent dissipation ε has been expressed as a function of the dissipation rate ω. The mutually exclusive activation of one among the two models is regulated by switches [Men94] estimating the dimensionless distance $y+$ [Bej84, p. 238]: depending on the position inside the flow either the k-ε model or the k-ω is solved. This model is known as the κ-ω SST model.

Once the equations for κ, $\varepsilon(\omega)$ and ω are solved, the kinematic eddy viscosity is derived from κ and ω as follows:

$$\nu_t = \frac{\kappa}{\omega}, \tag{4.24}$$

and the RANS equations can be solved.

The computed flow field is then used as an input to the solution of the energy conservation equation in a similar fashion to the laminar case.

Near wall treatment

The set of equations presented above requires further assumptions about the near wall treatment of the flow.

Preliminary and speculative considerations about the flow regime under the conditions in Tab. 4.2 led to the assessment of the flow being either laminar or turbulent ($Re \sim 1e04$) instead of fully turbulent ($Re \geq 1e05$). The near wall treatment chosen for the solution of the κ-ω SST model is known as Low-Reynolds (Low-*Re*) treatment. The main peculiarity of this approach is the use of damping functions to suppress the production of the turbulent quantities when the viscous dissipation due to μ_{He} is predominant.

The drawback of the Low-*Re* treatment is the need for highly dense computational grids close to the boundaries of the domain. Furthermore, it is required that the $y+$ [Sta08] value in the first cell departing from the wall shall be in the order of 1. More details about the computational grid generation will be given in the following section; here it is relevant to mention that one of the leading criteria for the mesh generation was having $y+ \sim 1$ in the first cell layer close to the wall. Nevertheless, the complexity of the meander flow geometry leads to an inhomogeneous helium flow. A computational grid where the first layer of cells had a constant height would not guarantee $y+ \sim 1$ over the entire domain. The generation of a suitable computational grid allowing $y+ \sim 1$ throughout the first layer of cells close to the wall requires a trial and error process. The repetition of this procedure for the meander flow geometry in Tab. 4.3, at each Re^* simulated, would have been extremely time-consuming; therefore a different solution has been searched. The aim was to reduce the overall analysis simulation time, while still keeping an adequate accuracy.

The Low-*Re* treatment has been therefore coupled with the so-called wall functions. These mathematical formulae are used to solve the velocity field close to the wall. The wall functions are very flexible in terms of required $y+$ value at the first cell close to the wall and less demanding regarding the mesh density in this region; though they are sufficiently accurate. These functions act as backup of the Low-*Re* treatment and are used instead of it in those regions of the computational domain where $y+ \neq 1$. This approach for the near wall treatment will be referred to as hybrid near wall treatment.

The resulting physical model for the simulation of the turbulent helium flow inside the meander flow geometry is the low-*Re* κ-ω SST model with hybrid wall treatment. The governing equations are described hereunder [Men94, Sta08, p. 2-18].

$$\rho_{\mathrm{He}} \cdot \left(\frac{d\kappa}{dt} + \boldsymbol{v} \bullet \nabla\kappa\right) = \tau_{ij} \cdot \nabla \bullet \boldsymbol{v} - \beta^* \cdot \rho_{\mathrm{He}} \cdot \omega \cdot \kappa + \nabla \bullet [(\mu_{\mathrm{He}} + \sigma_{\mathrm{k}} \cdot \mu_{\mathrm{t}}) \cdot \nabla\kappa], \tag{4.25}$$

$$\rho_{\mathrm{He}} \cdot \left(\frac{d\omega}{dt} + \boldsymbol{v} \bullet \nabla\omega\right) = \frac{\gamma}{\nu_{\mathrm{t}}} \cdot \tau_{ij} \cdot \nabla \bullet \boldsymbol{v} - \beta^* \cdot \rho_{\mathrm{He}} \cdot \omega^2 + \nabla \bullet [(\mu_{\mathrm{He}} + \sigma_{\omega} \cdot \mu_{\mathrm{t}}) \cdot \nabla\omega] + $$
$$+2 \cdot (1 - F_1) \cdot \rho_{\mathrm{He}} \cdot \sigma_{\omega 2} \cdot \frac{1}{\omega} \cdot \nabla\kappa \cdot \nabla\omega, \tag{4.26}$$

$$\phi = F_1 \cdot \phi_1 + (1 - F_1) \cdot \phi_2, \tag{4.27}$$

$$\nu_t = \frac{\kappa}{\omega} = \frac{a_1 \cdot \kappa}{\max(a_1 \cdot \omega, \Omega \cdot F_2)}, \tag{4.28}$$

$$F_2 = \tanh\left(\arg_2^2\right), \tag{4.29}$$

$$\arg_2 = \left(2 \cdot \frac{\sqrt{\kappa}}{0.09 \cdot \omega \cdot y}; \frac{500 \cdot \upsilon_{\mathrm{He}}}{y^2 \cdot \omega}\right). \tag{4.30}$$

Tab. 4.5: Coefficients for the set ϕ_1.

$\sigma_{\kappa 1}$	0.85
$\sigma_{\omega 1}$	0.5
β_1	0.0750
a_1	0.31
β^*	0.09
K	0.41
γ_1	$\frac{\beta_1}{\beta^*} - \frac{\sigma_{\omega 1} \cdot \kappa^2}{\sqrt{\beta^*}}$

Tab. 4.6: Coefficient for the set ϕ_2.

$\sigma_{\kappa 2}$	1.0
$\sigma_{\omega 2}$	0.856
β_2	0.0828
β^*	0.09
K	0.41
γ_2	$\frac{\beta_2}{\beta^*} - \frac{\sigma_{\omega 2} \cdot \kappa^2}{\sqrt{\beta^*}}$

4.2.6 Software, numerical method, mesh generation and solution algorithm

The CtFD analysis of the helium flow inside the meander flow geometry has been performed with two commercial software, namely Star-CCM+® and Star-CD®. Both software packages are produced by the company CD-Adapco. Star-CCM+® has been used to create the computational domain of the meander flow geometries listed in Tab. 4.3 and to generate the mesh used to discretize them. Afterwards, the discretized domains have been imported in Star-CD® and the boundary conditions, the material properties as well as the physical models to be solved have been set. Star-CD® has also been used to solve the models.

Star-CD® solves sets of Partial Differential Equations, PDEs, with the numerical method called Finite Volumes, FV. To apply this method, the computational domain has to be subdivided into a finite number of small control volumes. The computational grid defines therefore the boundaries of these control volumes, not the actual computational nodes. The set of PDEs is then converted into its integral formulation and applied to each control volumes inside the computational domain. The integrals are solved numerically. The FV method is particularly suitable for the thermal-fluid dynamic models because they are based on PDEs describing the conservation of some physical quantity. Indeed, global conservation is inbuilt into the FV method itself. Furthermore, the FV method can be easily applied to unstructured grids, which are the most suitable for the discretization of complicate geometries.

The computational grids for the meander flow geometries have been created with Star-CCM+®. The structure of the computational grids has to fulfil the requirements of the physical model to be solved. As stated in the previous section, the thermal-fluid dynamic analysis of the helium flow consists in solving the laminar flow model or the Low-*Re* κ-ω SST model with hybrid wall treatment. The solution of these two models requires computational grids that differ in particular in the mesh fineness close to the wall boundaries. However, the grids can be created from a common discretization concept. For this work, the computational domains have been discretized with unstructured grids. The grids were built with two different types of cells:

- Polyhedral cells,
- Prism cells.

Polyhedral cells are used to discretize the core flow region, whereas a suitable number of prism cell layers are used between the polyhedral cells and the wall boundaries. Details about the specific features of the laminar and turbulent grids are given below.

Typically, the numerical solution of a laminar flow problem does not require a particularly fine mesh close to the boundary; for this work, at least two layers of prisms cells have been used. The polyhedral cells have been entirely coated by homogeneously distributed prisms cells constituting the boundary layer. The mesh generator inbuilt in Star-CCM+® seeks the polyhedral cells to have on average 14 faces. The characteristic dimensions of the polyhedral cells range, mainly depending on the meander flow geometry and on the location, from 0.5 mm to 0.9 mm.

For the turbulent modelling the mesh in the boundary layer has to fulfil the requirements of the turbulence model in terms of height of the first layer departing from the wall (constraint on the $y+$ parameter), number of layers and prism height increment moving from the first layer towards the core flow region. For the present analysis, the boundary layer grid has the following characteristics:

- the height of the first layer is homogenous all over the computational domain and has been chosen to have $y+ \sim 1$, in agreement with the requirements of the Low-*Re* κ-ω SST model in Star-CD®; due to the complexity of the domain, locations where $y+ > 1$ occur and are therefore treated with the wall functions,

- in general ~ 15 prism layers have been used in the boundary layer,
- the height of each layer grows by a factor of 1.1 with respect to the previous one.

Regarding the polyhedral cells constituting the core mesh, they have on average 14 faces and their maximum size varies in the range 0.5 - 0.9 mm, mainly depending on the geometrical parameter *t* and on the location inside the computational domain. Examples of the mesh are shown in Fig. 4.5 and Fig 4.6: slices cutting the unstructured grid have been created in order to show its main features at different positions inside the computational domain. The computational results obtained on these kinds of mesh have been tested against grid independence, as shown in Appendix D.

The solution method used for the set of equations described above is a variant of the implicit algorithm SIMPLE [PS72] implemented in Star-CD®.

In conclusion of this section, the entire procedure of the CtFD analysis of the helium flow inside the meander flow geometry is summarized with the flow chart in Fig. 4.7.

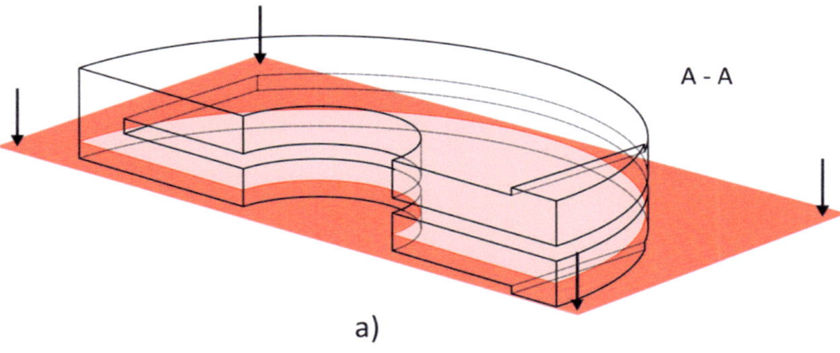

Computational mesh for the laminar regime model

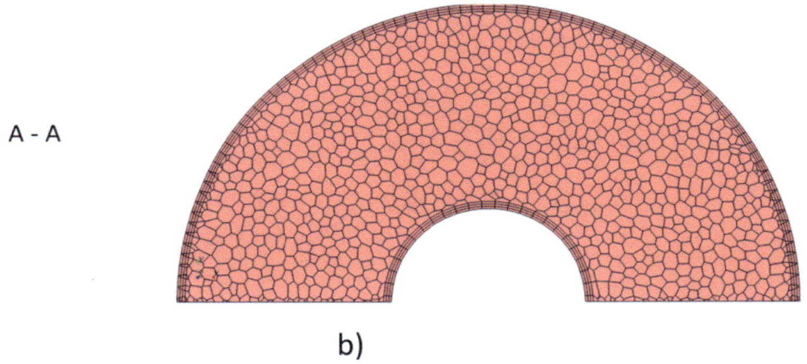

Fig. 4.5: a) Section cutting across the lower layer of the discretized computational domain;

b) Computational mesh for the laminar regime model. In this case only the mesh for the laminar flow model is shown because the focus is on unstructured polyhedral mesh characterizing the core flow region.

The mesh dimension is altered in order to increase the visibility of its main features.

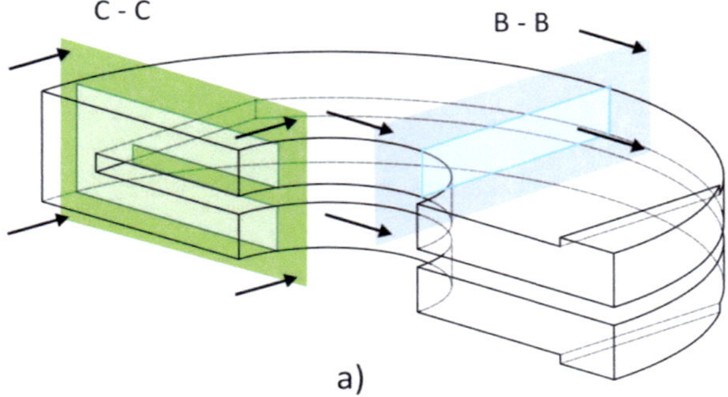

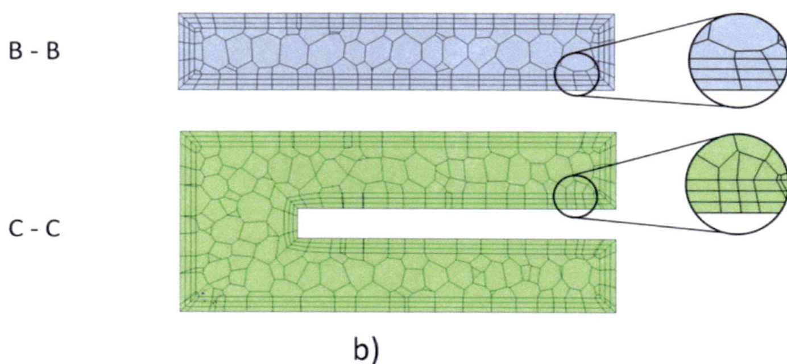

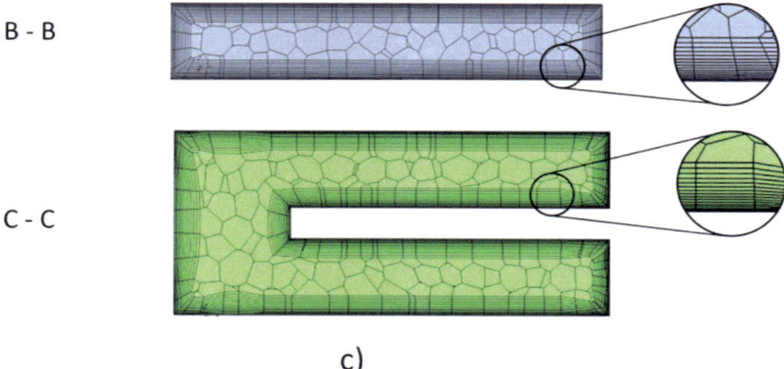

Fig. 4.6: a) Section of the upper layer and of the cut-off region of the discretized computational domain;

b) Computational mesh for the laminar regime model;

c) Computational mesh for the turbulent regime model.

The mesh dimension is altered in order to increase the visibility of its main features.

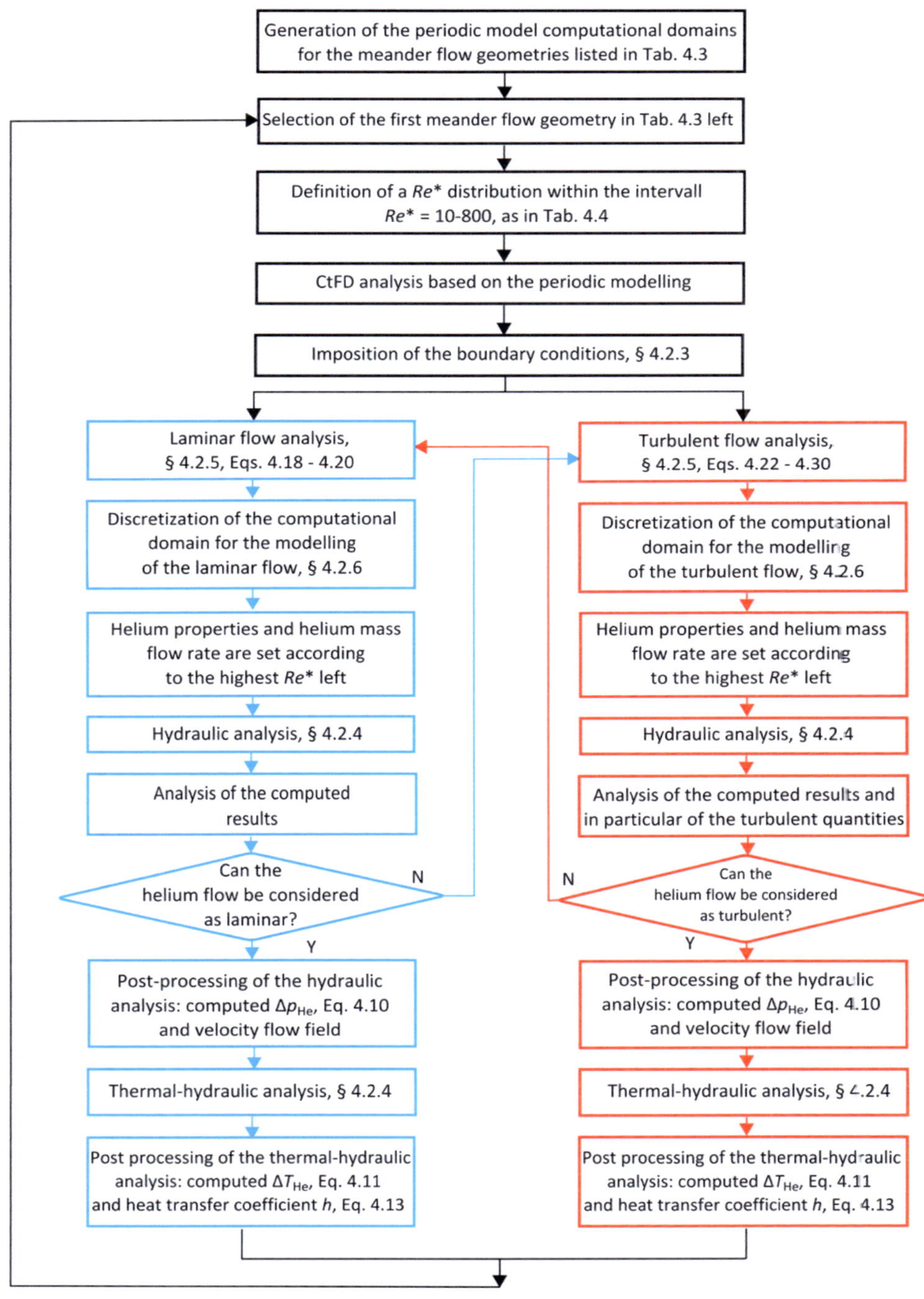

Fig. 4.7: Flowchart explaining the procedure the CtFD analysis with the periodic modelling.

4.3 Validation of the CtFD analysis based on the periodic modelling

The adoption of a novel computational strategy for the purpose of analysis and predictions must rely on a validation process. The expression *validation process* refers to the comparison among a set of experimental evidences and the computational results. The set of experimental evidences is necessarily limited; though it must be exhaustive and representative of the physics that the models implemented in the computational tool are meant to describe. Certainly the computational strategy has to be applied under analogous conditions as the experimental ones. In case the computational results successfully reproduce the experimental data, evidently within the experimental uncertainties, the computational strategy is said to be validated. Once it has been validated, the computational strategy can be confidently applied to a wider set of conditions, provided the physics of the phenomena to be modelled does not change.

The CtFD analysis based on the periodic modelling has to be validated as well; otherwise it would not be possible to rely on the results of the analysis of the helium flow inside the meander flow geometry. Firstly, a set of experimental results to be used in the validation procedure has to be defined.

Experimental database at KIT

The experimental campaign in which the thermal-hydraulics of the helium inside the meander flow geometry has been investigated consists basically of two phases.

During the first phase, short samples of meander flow heat exchanger (mock-ups) with different fin distances t have been tested. The goal was an assessment of the heat transfer performance as well as of the pressure drop in this geometry. A part of the campaign was also devoted to quantify possible buoyancy effects when the helium density gradient inside the heat exchanger is opposite to the helium flow direction: indeed, since the density of the helium decreases with the temperature, buoyancy effects in the warm side of the heat exchanger could drag the flow from the opposite side (i.e. from the cold side of the heat exchanger). Nevertheless it was found that the buoyancy effects are negligible, mainly due to the limited helium density ρ_{He} variation occurring over the temperature gradient 50 - 300 K (ρ_{He} varies by a factor of six)[5].

A detailed description of the experimental set-up as well as the experimental results can be found in [HL07] and [LHN08]; an overview on the meander flow geometry parameters of the heat exchangers as well as of the conditions under which they have been tested is shown in Tab. 4.7.

The second phase consisted in testing the W7-X HTS current lead prototypes mounting actual meander flow heat exchangers. The information about the experimental setup and results can be found in [FDF11, HDF11].

Tab. 4.7: Summary of the meander flow geometry mock-ups experimental conditions. Source [HCB10].

	Mock-up 1	**Mock-up 2**
Length, L / m	240	240
Outer diameter, d_o / mm	120	120
Central bar diameter, d_i / mm	40	40
Fin distance, t / mm	3	5
Fin thickness, s / mm	2	2
Cut off, c_o / mm	7	7
Helium mass flow rate / g/s	1.1 - 1.8	1.1 - 1.8
Helium inlet temperature / K	20 - 60	20 - 60
Helium inlet pressure / MPa	0.2 - 0.5	0.2 - 0.5
Mock-up cold temperature / K	~ 50 - 65	~ 50 -65
Mock-up warm temperature / K	~ 100	~ 100

The experimental results on the mock ups and the meander flow heat exchangers mounted in the W7-X HTS current lead prototype constitute the set of data for the validation of the CtFD analysis with the periodic modelling:

[5] This was of particular relevance for the operating conditions of W7-X; indeed, the HTS current leads will be mounted in a vertical upside-down arrangement, or with the cold side of the HTS current lead displaced above the warm one.

- the introduction of the periodic modelling and the first part of its validation were presented in [SCF10]. In this case, the periodic modelling has been shown to be capable of reproducing the experimental results of the meander flow heat exchanger mock-ups tested at the KIT,
- the second part of the validation involved the modelling of a full length meander flow heat exchanger mounted in an actual HTS current lead. The results of this second part have been presented in [RHS11]. It has been shown that with the results of the periodic modelling it was possible to reproduce the helium pressure drop occurring across the meander flow heat exchanger and the temperature profile along it (and the other components of the current leads).

Choice of the turbulence model

A computational tool solves numerically the equations of a specific physical model used to investigate a certain physical phenomenon. As mentioned above, it has to be validated against a set of experimental results, in order to be considered reliable.

The validation procedure becomes less straightforward when not only one, but rather several different models are available to describe the same physical phenomenon. Typically, when this situation occurs, the models have been derived from different assumptions; their range of applicability may vary as well as their capability in describing certain features of the phenomenon rather than others.

The choice of a suitable turbulence model to simulate a turbulent flow is an emblematic example of this problem. From the beginning of the twentieth century several approaches have been developed aiming at modelling and predicting the behaviour of turbulent flows. The literature about the turbulent modelling counts thousands of contributions. The complexity of these models has steadily increased due to the availability of more powerful computational resources. Nowadays, indeed, the computational approach to the turbulent analysis is the leading approach. The models typically implemented in the CtFD software available on the market can be classified as follows:

- Reynolds Average Navier-Stokes models, or RANS models:
  - Eddy viscosity models,
  - Reynolds stress transport models,
- Large Eddy Simulation models, or LES models,
- Detached Eddy Simulation models, or DES models,
- Direct Numerical Simulations, or DNS models.

The wide variety of turbulence models is due to several reasons. Firstly, none of them can catch completely the physics of the turbulence; their capability to predict certain phenomena occurring in a turbulent flow can differ significantly as well as their demand in terms of computational resources. A detailed description of the characteristics of each class of turbulence models is behind the scope of this work. This part rather focuses on the rationale behind the choice of the turbulent model described in § 4.2.5.

Within the class of RANS eddy viscosity models, the two-equation models have become more and more attractive in the last decades. This kind of models offers some relevant advantages. In the first place, these models are self-consistent, for the eddy viscosity is calculated from two turbulent quantities. For each turbulent quantity a conservation equation is solved. No further assumption on the scales of the turbulent flow is needed, as for the zero- and for the one-equation models. Furthermore they are less demanding in terms of computational power and CPU time than, for instance, the DNS models. Two-equation models have been widely and successfully applied to several types of flow; indeed, a huge literature is available about their performance. The eddy viscosity two-equation models have been adopted for the CtFD analysis with the periodic modelling.

Several eddy viscosity two-equation models are implemented in the software Star-CD®. The κ-ε model, the κ-ω model and the κ-ω SST model have been tested. Fig. 4.8 shows the comparison between the pressure drop computed with the CtFD periodic modelling and the experimental data. As it can be seen, the three turbulent models predict pressure drop values very close to each other. All of them underestimate the pressure drop with respect the experimental data by about 5%. This does not represent a drawback, since the experimental pressure drop data include contributions of the pressure drop occurring into the feeding pipes (due to some constrains in the experimental set up it cannot be avoided) as well as the contribution due to the inlet and outlet effects (which cannot be modelled with the CtFD periodic modelling). It is therefore reasonable that the computed result underestimate the pressure drop.

For the modelling of turbulent flows with the CtFD periodic modelling, the κ-ω SST has been adopted. It shows a good reliability in reproducing the experimental results and it takes advantage of the characteristics of the κ-ω and the κ-ε to simulate the different regions of a turbulent flow.

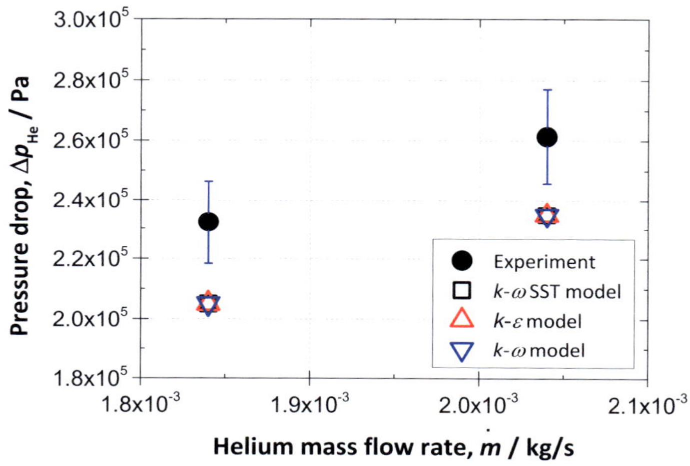

Fig. 4.8: Comparison between the experimental and the computed pressure drop. Three different turbulent models have been tested.

4.4 Summary

In this Chapter, the objective of the thermal-fluid dynamic analysis of the helium flow inside the meander flow geometry has been presented in more detail. The goal is to relate the heat transfer coefficient h and the pressure drop coefficient ζ to a dimensionless quantity characterizing the helium flow: the Reynolds number, Re. This requires the analysis of the helium flow in several, different meander flow geometries and under different flow conditions. A set of twenty-five different meander flow geometries, the material properties and the mass flow rate have been set according to the requirements of the HTS current leads presently under design or construction. The analysis itself has been performed with a CtFD technique based on the periodic modelling of small portions of the helium channel in the meander flow heat exchanger. Firstly, the analysis shall provide a guideline to discern the flow regime over the range of flow conditions covered in this work, i.e. if the flow regime is laminar or rather turbulent. Then, the heat transfer coefficient h and the pressure drop coefficient ζ have to be derived for all meander flow geometries, at every flow condition investigated.

5 Correlations for the helium thermal-fluid dynamics in the meander flow geometry

In this Chapter, the results of the CtFD periodic modelling are described in detail. In the first place, definitions for the geometrical quantities characterizing the helium flow in the meander flow geometry, i.e d_h and A_{He}, are presented. The helium laminar and turbulent flow regime are then characterized depending on the Reynolds number *Re*. In conclusion, correlations for the pressure drop coefficient ζ and the Nusselt number *Nu* for both the laminar and the turbulent regime are derived. The content of this Chapter has been published by the author in [RHS13a, RHS13b].

5.1 Characteristic geometrical quantities of the meander flow geometry

In the previous Chapter it has been shown that the helium flow conditions inside the meander flow heat exchanger are characterized by the Reynolds number *Re*. As recalled in Eq. 5.1, the Reynolds number depends on the mass flow rate $\dot{m}$, on the molecular viscosity μ_{He} (therefore on the thermodynamic conditions of the helium) and on the ratio d_h/A_{He}.

$$Re = \frac{\dot{m}}{\mu_{He}(p_{He}, T_{He})} \cdot \frac{d_h}{A_{He}}. \tag{5.1}$$

The two quantities d_h and A_{He} are called hydraulic diameter and helium flow cross section, respectively. The first one has the dimension of a length, whereas the second of an area. Both these quantities are not known a priori, since their formulation depends on how the helium flows inside the meander flow geometry. For this reason, the dimensional quantity *Re** has been introduced in the previous Chapter in order to allow a more comfortable handling of the mass flow rate and the thermodynamic conditions.

The relation among the Reynolds number *Re* and *Re** is recalled in Eq. 5.2:

$$Re^* = Re \cdot \frac{A_{\mathrm{He}}}{d_{\mathrm{h}}} = \frac{\dot{m}}{\mu_{\mathrm{He}}(p_{\mathrm{He}}, T_{\mathrm{He}})}. \tag{5.2}$$

For a specific meander flow geometry, Re^* provides information on the flow analogous to Re, but geometry-dependent.

The first part of the post-processing of the CtFD analysis results deals with the definition of the hydraulic diameter d_{h} and the helium flow cross section A_{He} in the meander flow geometry. As stated in [Web94], the definition of such geometrical quantities for complicate geometries is not trivial and requires the analysis of the fluid flow in turbulent regime. The helium flow inside all meander flow geometries listed in Tab 4.3 has been therefore analysed at the corresponding largest Re^* value, or at $Re^* > 700$. Indeed, as it will be shown later on in this Chapter, at $Re^* > 700$, the helium flow regime is turbulent in all meander geometries considered in this work.

Firstly, the helium flow cross section A_{He} is defined.

5.1.1 Characteristic helium flow cross section, A_{He}

The helium flow inside the meander flow geometry experiences a continuous and periodic change of cross section, as shown in Fig. 5.1 a). Depending on the outer diameter d_{o}, on the central bar diameter d_{i} and on the cut off c_{o}, two behaviours are possible; this is shown in Fig. 5.1 b), where the dependence of the helium cross section A_{He} on the fictitious coordinate ξ has been plotted. In the first case (dashed-dotted black line), the helium cross section at the beginning and at the end of the region between two fins, named A_0, is smaller than the helium cross section in the middle of the channel, where $A_{\mathrm{He}} = (d_{\mathrm{o}} - d_{\mathrm{i}}) \cdot t$. In the second case (dotted blue line), the helium cross section A_0 is larger than the helium cross section in the middle of the channel. Though, the helium cross section A_{He} becomes larger in the region directly after A_0, before starting to shrink in correspondence of the central bar.

The definition of the helium flow cross section characterizing the helium flow aims at providing guidelines to evaluate a representative value of A_{He}, given a specific meander flow geometry defined by d_{o}, d_{i}, t, s, and c_{o}.

To achieve this goal, the helium flow in the turbulent regime has been analysed on a plane cutting at mid-height the region between the fins. As mentioned above, two cases have been considered: in the first case $A_0 < (d_o - d_i) \cdot t$; therefore the ratio $(A_0 / (d_o - d_i) \cdot t)$ is smaller than one. In the second case $A_0 > (d_o - \mathrm{di}) \cdot t$; therefore the ratio $(A_0 / (d_o - d_i) \cdot t)$ is larger than one. The results for three meander flow geometries with the ratio $(A_0 / (d_o - d_i) \cdot t)$ increasing towards one are shown in Fig. 5.2. The results for three meander flow geometries with the ratio $(A_0 / (d_o - d_i) \cdot t)$ departing from one are shown in Fig. 5.3.

When the ratio $(A_0 / (d_o - d_i) \cdot t)$ is smaller than one, the helium flowing into the region between the fins experiences an intense adverse pressure gradient. Two factors contribute to the adverse pressure gradient: as it has been shown in Fig. 5.1 b), the helium cross section always increases directly after A_0. Consequently the flow velocity slows down directly after A_0 and a local adverse pressure gradient is generated. Moving further towards the middle of the channel, the velocity of the helium remains lower than at A_0, since the cross section A_0 is smaller than $(d_o - d_i) \cdot t$; therefore the resultant net pressure gradient acts opposite to the flow direction. Consequently, the inflowing helium stream detaches from the external wall and recirculation regions, or vortices, are generated. The extension of the recirculation region decreases as the ratio $(A_0 / (d_o - d_i) \cdot t)$ increases, as it can be clearly appreciated on the right side of Fig. 5.2. Less extended vortices are generated by the detachment of the helium flow from the central bar.

In the meander flow geometries having the ratio $(A_0 / (d_o - d_i) \cdot t)$ smaller than one, the presence of large recirculation regions delimits a main flow channel whose cross section basically corresponds to A_0. The main flow channel corresponds to the region where the helium flow velocity is higher, as it can be clearly seen in Fig. 5.2.

When the ratio $(A_0 / (d_o - d_i) \cdot t)$ is larger than one, the helium flowing into the region between the fins only experiences a local adverse pressure gradient. Indeed, since the helium cross section increases directly after A_0, the flow velocity slows down and a local adverse pressure gradient is generated. Moving further towards the middle of the channel, the velocity of the helium increases with respect the velocity at A_0, since the cross section A_0 is larger than $(d_o - d_i) \cdot t$; as a consequence of the local adverse pressure gradient, the inflowing helium stream detaches from the external wall directly after the inlet and small vortices are generated.

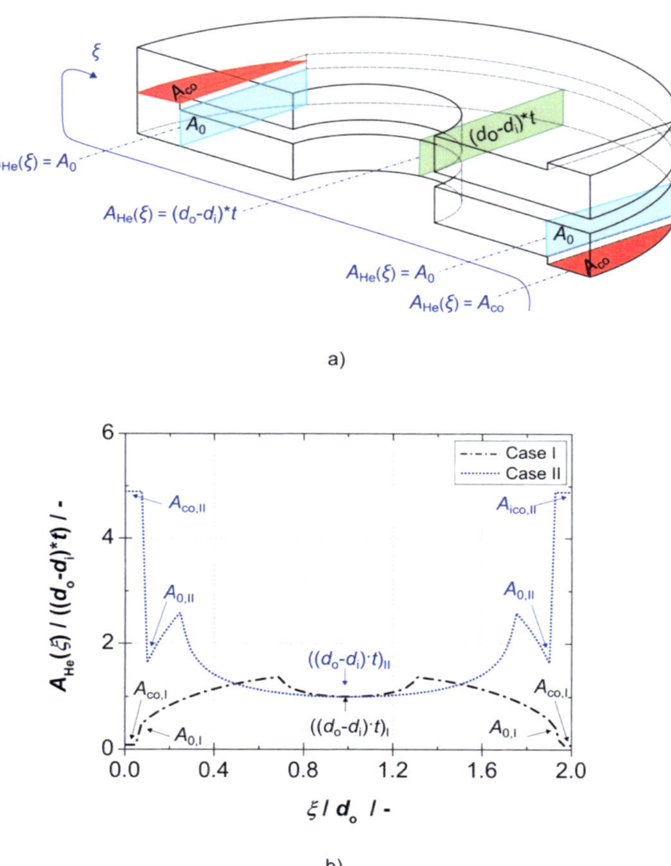

Fig. 5.1: a) Qualitative representation of the helium flow cross section A_{He} inside the meander flow geometry as a function of the fictitious coordinate ξ;

b) Behaviour of the helium cross section A_{He} as a function of ξ for the case $A_0 < (d_o - d_i)\cdot t$ (Case I) and $A_0 > (d_o - d_i)\cdot t$ (Case II). To ease the comparison, the coordinate ξ has been normalized to the outer diameter d_o, whereas A_{He} has been normalized to $(d_o - d_i)\cdot t$.

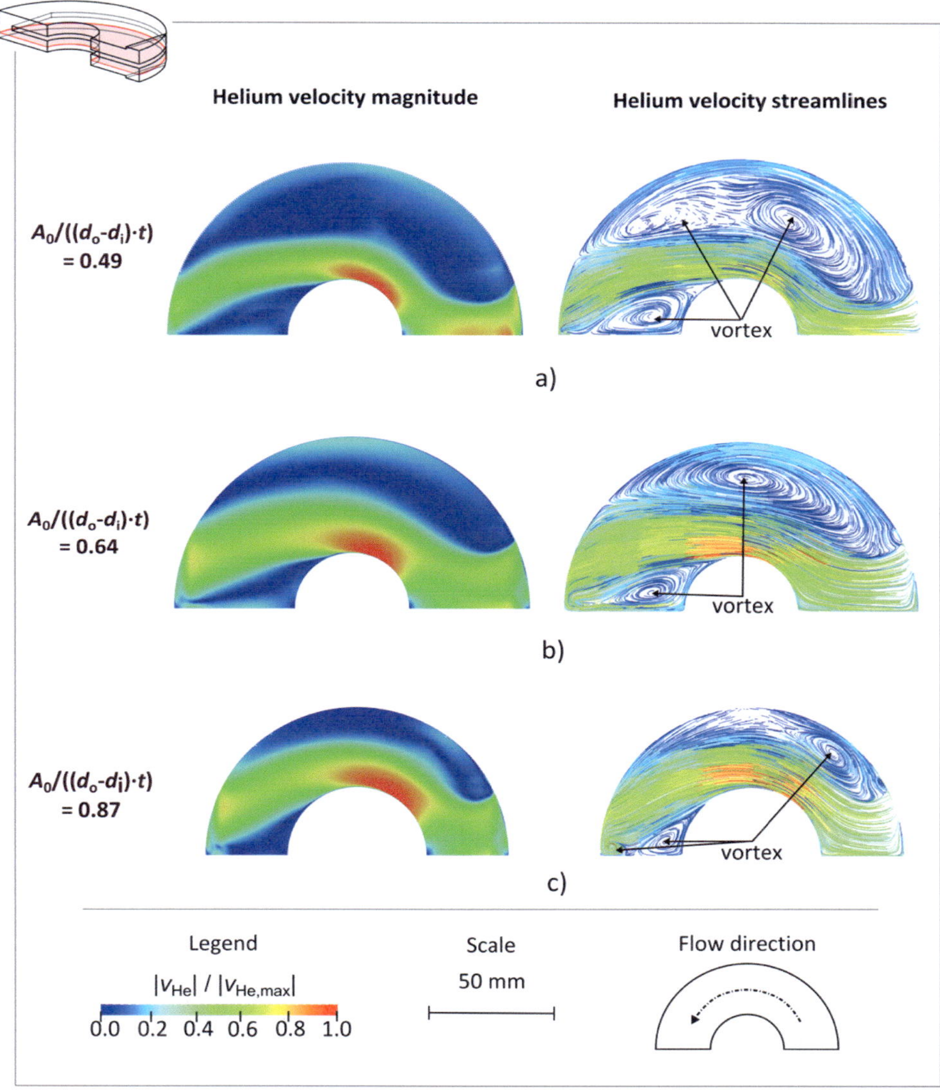

Fig. 5.2: Turbulent helium flow between the fins ($Re^* > 700$), $A_0 < (d_o - d_i) \cdot t$:

a) Macro group IV, meander flow geometry #22, Tab 4.3
d_o = 140 mm, d_i = 45 mm, t = 5 mm, s = 2 mm, c_o = 4 mm;

b) Macro group II, meander flow geometry #13, Tab 4.3
d_o = 140 mm, d_i = 45 mm, t = 5 mm, s = 2 mm, c_o = 7 mm;

c) Macro group I, meander flow geometry #2, Tab 4.3
d_o = 120 mm, d_i = 55 mm, t = 3 mm, s = 2 mm, c_o = 7 mm.

The extension of the recirculation region decreases as the ratio $(A_0 / (d_o - d_i)\cdot t)$ increases, as shown on the right side of Fig. 5.3. As for the previous case, less extended vortices are generated by the detachment of the helium flow from the central bar.

If the ratio $(A_0 / (d_o - d_i)\cdot t)$ is larger than one, a helium main flow channel still exists whose cross section approximately corresponds to $(d_o - d_i)\cdot t$. This approximation becomes more precise as the ratio $(A_0 / (d_o - d_i)\cdot t)$ becomes larger.

According to this analysis, the helium cross section A_{He} characterizing the helium flow in the meander flow geometry is defined by choosing the minimum among the quantities A_0 and $(d_o - d_i)\cdot t$, as shown in Eq. 5.3.

$$A_{He} = \min(A_0, (d_o - d_i) \cdot t). \tag{5.3}$$

The value of A_0 can be calculated from the outer diameter d_o, the fin distance t and the cut-off width c_o as follows:

$$A_0 = 2 \cdot \left[\left(\frac{d_o}{2}\right)^2 - \left(\frac{d_o}{2} - c_o\right)^2 \right]^{0.5} \cdot t. \tag{5.4}$$

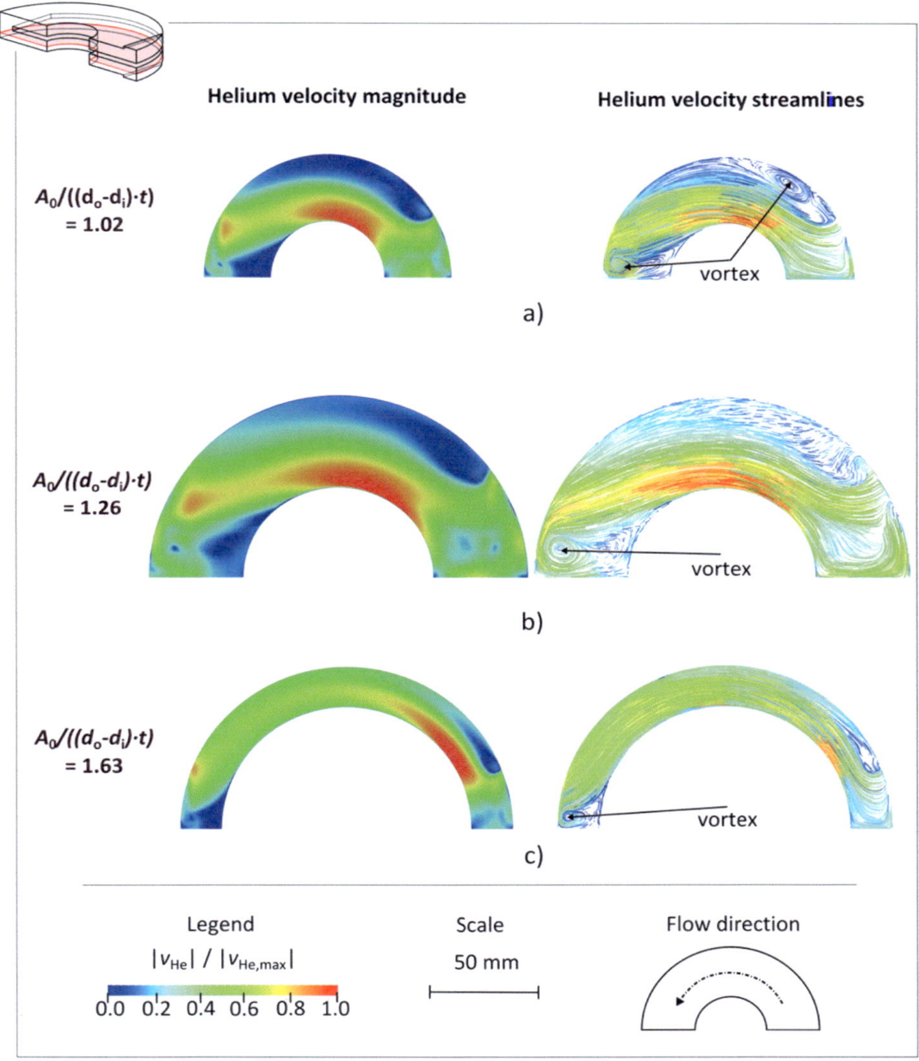

Fig. 5.3: Turbulent helium flow between the fins ($Re^* > 700$), $A_0 > (d_o - d_i)\cdot t$:

a) Macro group IV, meander flow geometry #21, Tab 4.3
d_o = 120 mm, d_i = 55 mm, t = 3 mm, s = 2 mm, c_o = 10 mm;

b) Macro group II, meander flow geometry #16, Tab 4.3
d_o = 180 mm, d_i = 90 mm, t = 8 mm, s = 2 mm, c_o = 20 mm;

c) Macro group I, meander flow geometry #9, Tab 4.3
d_o = 160 mm, d_i = 120 mm, t = 3 mm, s = 2 mm, c_o = 7 mm.

5.1.2 Hydraulic diameter, d_h

The definition of the hydraulic diameter d_h in complicated geometries as the meander flow geometry is not trivial. This quantity refers to a geometrical parameter, namely a length, characterizing the velocity of the flow inside the channel. In the previous section it has been shown that, in the turbulent regime, a main flow channel is originated between the fins of the heat exchanger. This main flow channel has a cross section A_{He} that depends on the geometrical parameters of the specific meander flow geometry. Indeed, the cross section A_{He} can be estimated to a good approximation as the minimum among the cross section A_0 and $(d_o - d_i) \cdot t$. In both cases, the helium flow cross section A_{He} has a rectangular shape. The width a of the helium cross section A_{He} is:

$$a = 2 \cdot \left[\left(\frac{d_o}{2} \right)^2 - \left(\frac{d_o}{2} - c_o \right)^2 \right]^{0.5}, \tag{5.5}$$

in the case $A_{He} = A_0$ and

$$a = (d_o - d_i), \tag{5.6}$$

in the case $A_{He} = (d_o - d_i) \cdot t$.

The height of the helium cross section is always given by the fin distance t. Since the fin distance t is typically one order of magnitude smaller than the width a (this can be easily seen in Tab. 4.3), the main flow channel inside the heat exchanger can be seen, to a certain extent, as a slender rectangular duct. The hydraulic diameter of this trivial geometry depends on the height and on the aspect ratio of the channel [Bej84, pag. 80]. To understand how the helium flow behaves inside the main flow channel discussed above, it is necessary to analyse the vertical velocity distribution. The analysis of the vertical velocity distribution in the meander flow geometry aims at:

- verifying if a stable vertical velocity profile inside the main flow channel can exist,
- if it does, comparing it with a typical turbulent profile [KSA87, p. **4**•62].

The velocity profile has been analysed inside the main flow channel in meander flow geometries with several values of the fin distance t.

The results are shown in Fig. 5.4 for the case with $t = 3$ mm, in Fig. 5.5 for the case with $t = 5$ mm and in Fig. 5.6 for the case with $t = 8$ mm. In all cases, the vertical velocity profiles have been plotted at three positions along the main flow channel, from the inlet towards the outlet of one layer of the computational domain. In order to generalize the approach, the velocity has been normalized to its maximum value. The vertical coordinate has been normalized to the specific fin distance t in a local coordinate system having the origin at a quote equal to $t/2$. As it is possible to see in Fig. 5.4, Fig. 5.5 and Fig. 5.6, the velocity profiles are always perturbed after the inlet region. Nevertheless, moving further towards the outlet of the layer, the velocity profile stabilizes and assumes a typical turbulent flow shape [KSA87, p. **4•**62];

According to these results, it is justified to define the hydraulic diameter d_h in the meander flow geometry approximating the mean flow channel to the flow in a slender rectangular duct. Since the cross section A_{He} of the main flow channel is known from Eqs. 5.3 – 5.4, as it is the width a (from Eqs. 5.5 – 5.6) and its height t, the hydraulic diameter d_h in the meander flow geometry can be calculated as:

$$d_h = \frac{2 \cdot A_{He}}{a + t} \sim 2 \cdot t, \tag{5.7}$$

where the approximation $d_h = 2 \cdot t$ holds if the width a is consistently larger than the fin distance t.

With Eqs. 5.3 – 5.4 and Eq. 5.7, it is now possible to convert the dimensional parameter Re^* to the dimensionless Reynolds number Re. According to the set of helium flow conditions listed in Tab. 4.2 and the meander flow geometries listed in Tab. 4.3, the Reynolds number range covered in the present work spans the interval $Re = 250$ - 30000.

From now on, the helium flow conditions will be referred to via the corresponding Reynolds number Re and no longer via the parameter Re^*.

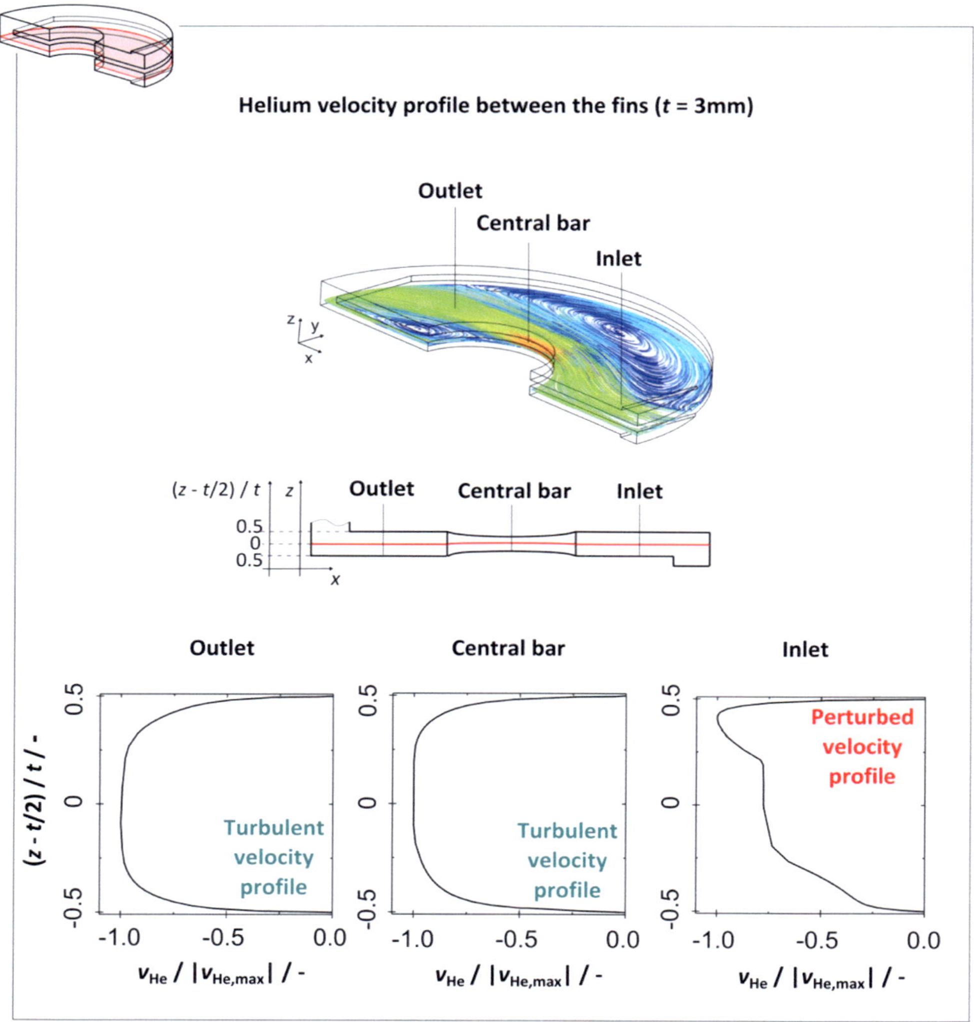

Fig. 5.4: Turbulent velocity profile ($Re^* > 700$), fin distance $t = 3$ mm;

Macro group I, meander flow geometry #4, Tab 4.3
$d_o = 140$ mm, $d_i = 45$ mm, $t = 3$ mm, $s = 2$ mm, $c_o = 7$ mm;

The velocity profile is shown at three positions inside the main flow channel: inlet, central bar and outlet. Moving away from the inlet region, where the velocity profile is perturbed, helium flow is stable and has the flat velocity distribution typical of the turbulent flow.

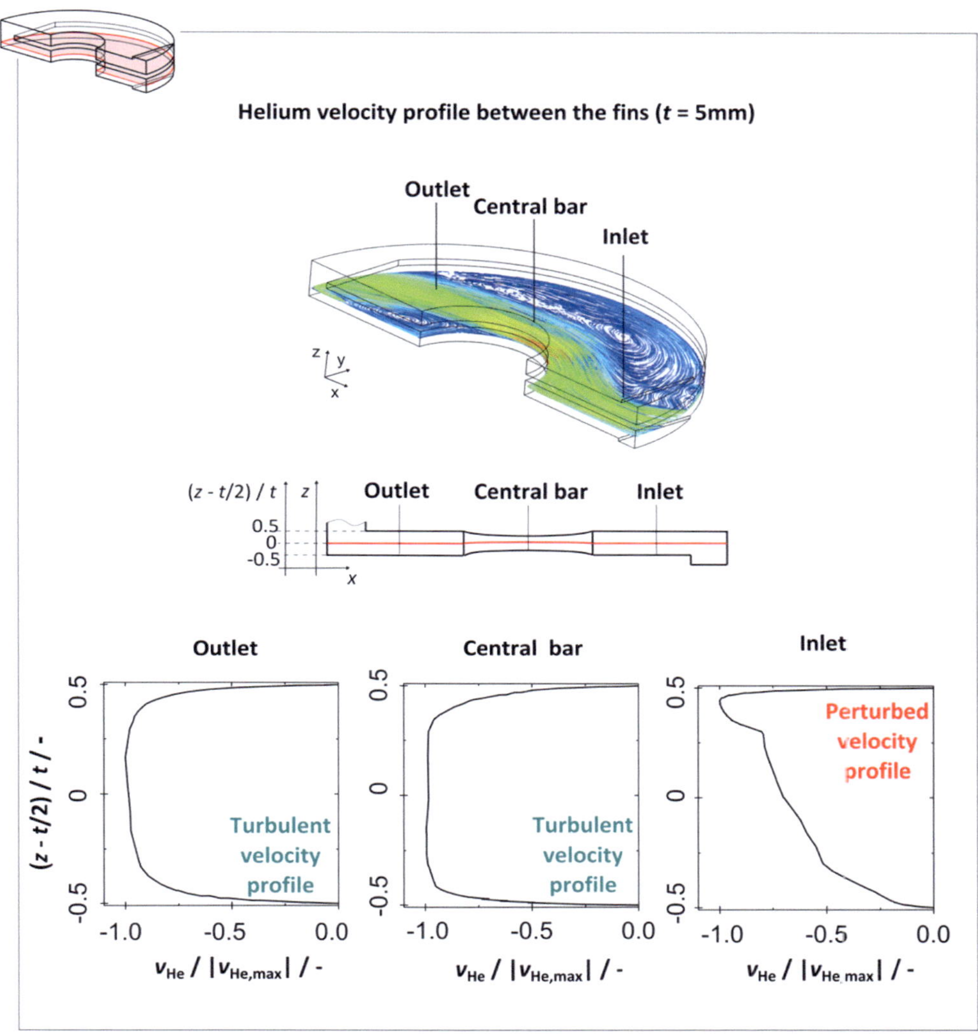

Fig. 5.5: Turbulent velocity profile (Re^* > 700), fin distance t = 5 mm;

Macro group II, meander flow geometry #13, Tab 4.3
d_o = 140 mm, d_i = 45 mm, t = 5 mm, s = 2 mm, c_o = 7 mm;

The velocity profile is shown at three positions inside the main flow channel: inlet, central bar and outlet. Also in this case, the velocity profile is perturbed close to inlet region. In the central bar and in the outlet region, the velocity profile of the helium flow is stable and has the flat distribution typical of the turbulent flow.

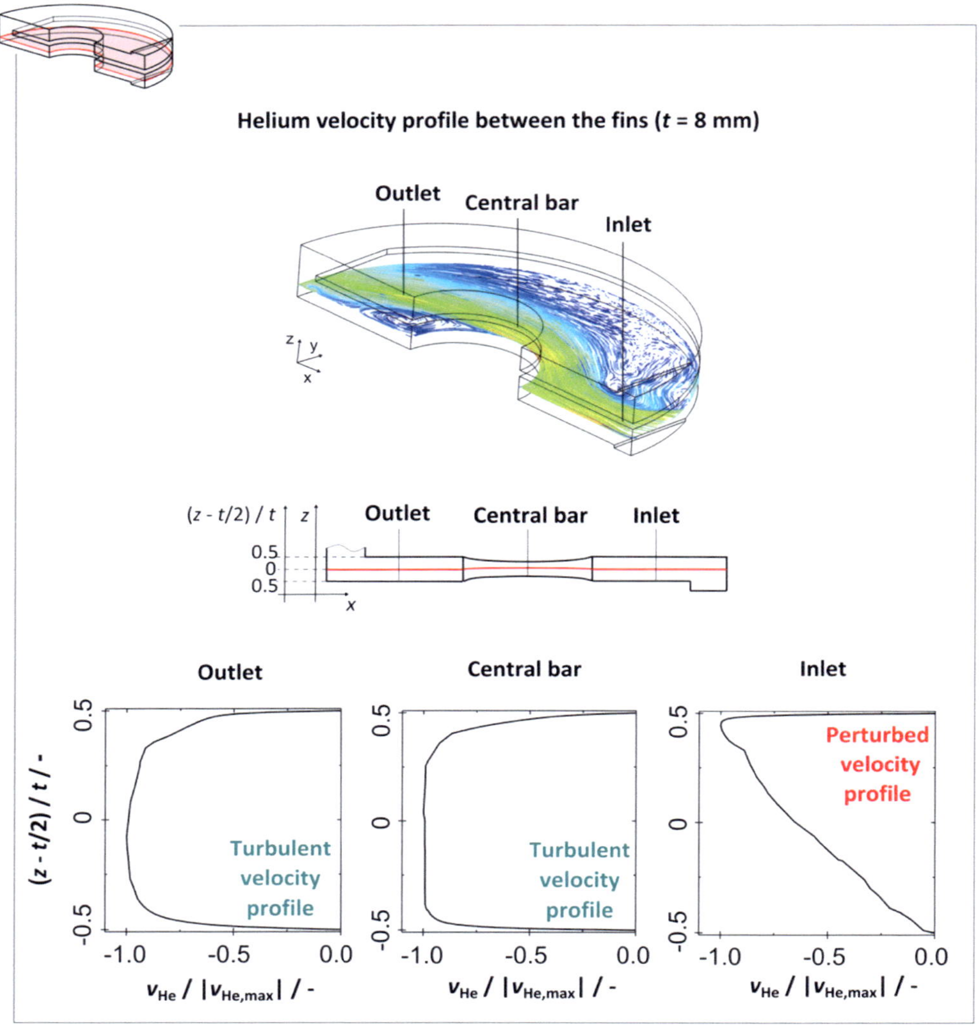

Fig. 5.6: Turbulent velocity profile ($Re^* > 700$), fin distance $t = 8$ mm;

Macro group II, meander flow geometry #14, Tab 4.3
$d_o = 140$ mm, $d_i = 45$ mm, $t = 8$ mm, $s = 2$ mm, $c_o = 7$ mm;

The velocity profile is shown at three positions inside the main flow channel: inlet, central bar and outlet. As for the cases with fin distances $t = 3$ mm and $t = 5$ mm, the velocity profile is perturbed close to inlet region. In the central bar and in the outlet region, the velocity profile of the helium flow is stable and has the flat distribution typical of the turbulent flow.

5.2 Flow regime inside the meander flow geometry

In the previous Chapter it has been explained that the helium flow regime, i.e. laminar or turbulent regime, is not known a priori in the meander flow geometry. This means that, assuming a certain helium flow condition in a meander flow geometry has been provided and it is expressed via the Reynolds number *Re*, it is not possible to state if the helium flow regime is laminar or turbulent.

One basic goal of the CtFD analysis is the assessment of the flow regime inside the meander flow geometry. For this reason, both the laminar modelling and the turbulent modelling based on the κ-ω SST model have been applied to the study of helium thermal-fluid dynamics. For each meander flow geometry listed in Tab. 4.3, the turbulent modelling has been applied starting from the largest Reynolds number *Re* towards the lower ones. On the contrary, the laminar modelling has been applied starting from the lowest Reynolds number *Re* towards the largest.

In this section, the procedure adopted to characterize the helium flow regime depending on the Reynolds number *Re* is presented.

5.2.1 Definition of the turbulent regime

The turbulent model named κ-ω SST model has been extensively discussed in § 4.2.5. This turbulent model belongs to the class of the RANS eddy viscosity models. The turbulent flow is described with the introduction of the quantity eddy viscosity μ_t, which comes out from the Reynolds averaging of the Navier-Stokes equations.

The eddy viscosity μ_t originates an extra dissipative term in the Navier-Stokes equations, which sums up to the molecular viscosity of the fluid μ_{He} (see Eqs. 4.21 - 4.23).

In fully turbulent flows, the eddy viscosity μ_t is typically at least two orders of magnitude larger than the molecular viscosity μ_{He}. On the other hand, laminar flows are characterized by a vanishing turbulent viscosity μ_t. Against this background, the turbulent viscosity μ_t has been systematically compared with the molecule viscosity for decreasing Reynolds numbers. The results of this comparison show that, for all cases considered, the computed turbulent viscosity μ_t is larger than the molecular viscosity at high Reynolds numbers *Re*. The magnitude of μ_t becomes comparable to that of μ_{He} at $Re \approx 2000$.

For helium flows with a Reynolds number lower than $Re \approx 2000$, the turbulent viscosity μ_t rapidly sinks to a few percentage of the molecular viscosity μ_{He}.

An example is shown in Fig. 5.7. For the sake of generality, the turbulent viscosity has been normalized with respect to the molecular viscosity. The distribution of the ratio μ_t/μ_{He} has been plotted on the mid-plane cutting one layer of the computational domain, at $Re \leq 1500$ and $Re \approx 2000$. Three cases have been considered, which are of general interest for characterizing the turbulent regime in the meander flow geometry. Indeed, the three meander geometries shown in Fig. 5.7 differ in terms of ratios of the cut-off c_o and the central bar diameter d_i to the other meander flow geometry parameters, i.e. the outer diameter d_o, fin distance t and fin thickness s.

The interaction of the helium flow with the central bar and the flow in the cut-off region clearly represent the main candidates to trigger instabilities in the flow itself and therefore a turbulent behaviour. The consequences in terms of turbulent behaviour of both these interactions can be appreciated in the flow between the plates. It can be clearly seen that the turbulent viscosity tent to vanish for Reynolds numbers lower than $Re < 2000$. Also in the last case, characterized by a very small cut-off c_o in comparison to the other geometrical parameters, at $Re \leq 1500$ the turbulent viscosity in the main flow channel is about two orders of magnitude lower than the molecular viscosity.

According to the considerations listed above, the incipit for the turbulent regime in the meander flow geometry has been set at $Re \approx 2000$. For the present work, the turbulent regime spans over the range $Re = 2000$ - 30000.

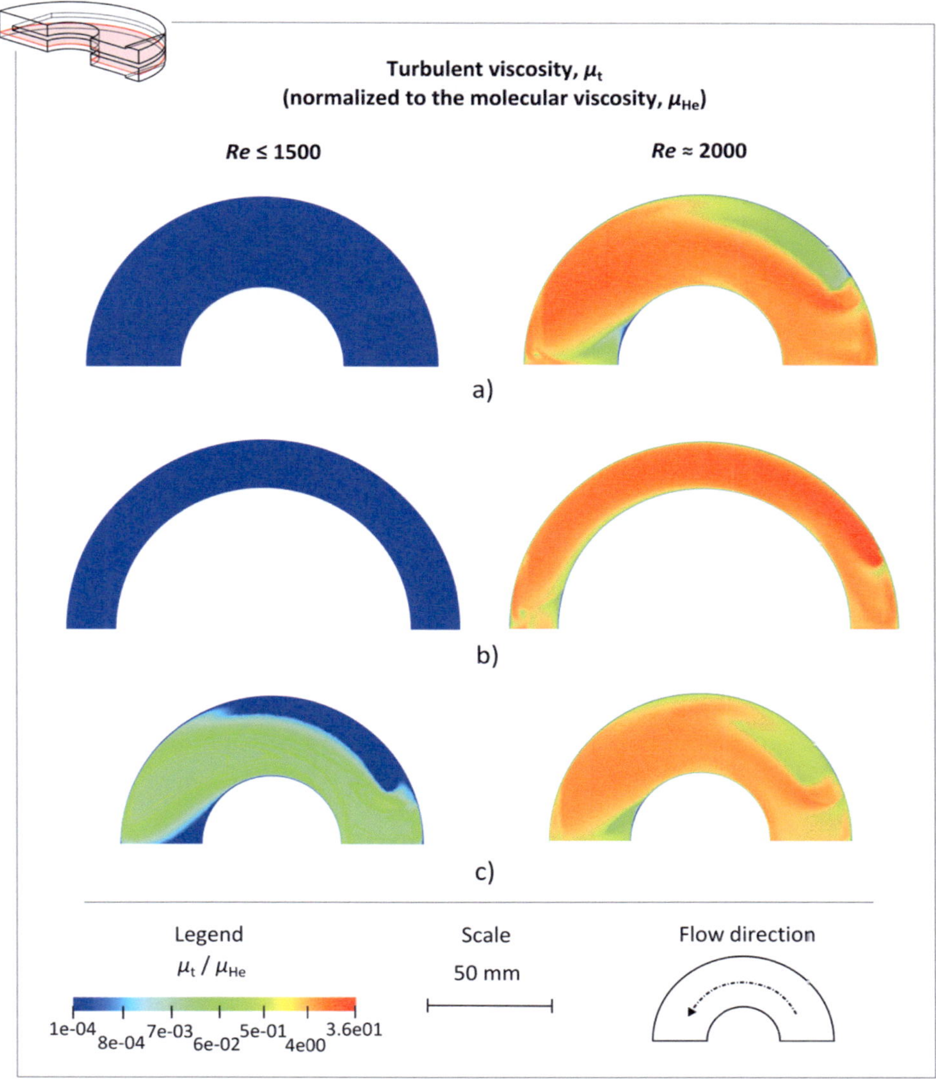

Fig. 5.7: Turbulent viscosity μ_t distribution between the fins at $Re \leq 1500$ and $Re \approx 2000$ (normalized to μ_{He}):

a) Macro group IV, meander flow geometry #23, Tab 4.3
$d_o = 140$ mm, $d_i = 45$ mm, $t = 5$ mm, $s = 2$ mm, $c_o = 10$ mm;

b) Macro group I, meander flow geometry #9, Tab 4.3
$d_o = 160$ mm, $d_i = 120$ mm, $t = 3$ mm, $s = 2$ mm, $c_o = 7$ mm;

c) Macro group IV, meander flow geometry #20, Tab 4.3
$d_o = 120$ mm, $d_i = 55$ mm, $t = 3$ mm, $s = 2$ mm, $c_o = 4$ mm.

5.2.2 Definition of the laminar regime

The definition of the laminar regime in the meander flow geometry is not trivially complementary to the definition of the turbulent regime. Indeed, it would be expected that some transition regime region may exist in this geometry as well, as it does in simpler ones [Bej84, p. 202]. This means that, although the turbulent regime has been defined in the range *Re* = 2000 - 30000, the laminar regime cannot be automatically defined in the range *Re* = 250 - 2000. Important further considerations are needed.

The laminar regime can be seen as an "anomaly" of the fluid flows relevant for technical applications, which are in general turbulent. For internal flows, its existence is guaranteed only under strict conditions involving the geometry and the fluid velocity along with the fluid thermodynamic properties. At a first glance, the existence of the laminar regime in the meander flow geometry does not appear obvious, even at low Reynolds numbers. Nevertheless, the analysis of the turbulent regime shows that the order of magnitude of the turbulent viscosity μ_t becomes comparable to that of the molecular viscosity μ_{He} at $Re \approx 2000$. If one attempts to extend the turbulent modelling to lower Reynolds numbers *Re*, a strong decrease of the computed μ_t (up to four orders of magnitudes lower than μ_{He}) is noted over the entire domain, in all the meander flow geometries. This is due to the strong damping of the turbulent quantities and in particular of the turbulent kinetic energy κ, which is directly proportional to μ_t (see Eq. 4.23). For the turbulence model itself predicts the turbulent contribution to be negligible, its applicability below $Re \approx 2000$ is questionable, although no numerical issues have been encountered during the solution.

Justifying the extension of the turbulent modelling to $Re < 2000$ as an attempt to investigate some features of the transition regime region is also arguable. Indeed, the commonly accepted capability of the low-*Re* models to capture some of the effects of the transition [Wil94] can lead, for non-trivial geometries, to the simulation of a "pseudo-transition" [MLV06] rather than the actual, physical change of flow regime. For a more reliable modelling of the transition, the low-*Re* turbulent models have to be coupled with more complex tools, as shown for instance in [MLV06]. These tools are not provided within the code used for the present work and the Star-CD®

documentation firmly advises against modelling (potentially) transitional flows with any of the turbulence models implemented therein [Sta08], for none of them can accurately model this flow regime.

The CtFD analysis based on the laminar modelling has been therefore applied to study the helium thermal-fluid mechanics in the meander flow geometry in case $Re < 2000$. The laminar modelling has been applied to the meander flow geometries models listed in Tab. 4.3 from the lowest Re values towards the larger ones.

One of the most relevant characteristics of the laminar regime is the direct proportionality between the mass flow rate and the associated pressure drop. Indeed, for simple geometries (i.e. circular ducts) the Navier-Stokes equations, under the assumption of fully developed flow and null inertia forces, reduce to a Poisson-type equation, whose analytical solution leads to the well-known Hagen-Poiseuille velocity profile. The derived friction factor scales then with $1/Re$ [Bej84, p. 78]. Regarding the laminar flow in the meander flow geometry, the simulations show that, if the computed pressure drop is arranged in terms of pressure drop coefficient ζ

$$\zeta = 2 \cdot \Delta p \cdot \rho_{\mathrm{He}}(p_{\mathrm{He}}, T_{\mathrm{He}}) \cdot \left(\frac{\dot{m}}{A_{\mathrm{He}}}\right)^{-2}, \tag{5.8}$$

it scales with $(1/Re)^n$ for Reynolds numbers up to $Re \approx 1000$, and $n \approx 0.52$.

Departing from $Re \approx 1000$, the relation between the pressure drop coefficient ζ and the Reynolds number Re changes and it is normally associated either to a lower reduction rate or to an asymptotic behaviour of the numerical residuals. Although a flow transition region is supposed to exist in this geometry as well, the laminar model cannot trigger it (nor can the turbulent model, as explained above).

Considering therefore the results of the CtFD analysis and the lack of experimental results on a local scale (i.e. on the scale of a period of the meander flow geometry) at $Re \approx 1000$, the boundary $Re = 1000$ has been adopted as upper limit of the range that will be referred to as laminar regime. Indeed, the behaviour of ζ for $Re < 1000$ suggests that, under these flow conditions, the relation between the computed flow field and the pressure drop does not change [RHS13b]. Consequently, the range $Re = 1000 - 2000$ is referred to as the transition regime region although the lack of appropriate models does not allow a more precise investigation. A possible treatment for helium flows at $Re = 1000 - 2000$ is discussed in § 7.2.2.

5.3 Correlation for the pressure drop coefficient, ζ

Once criteria to characterize the flow regime in the meander geometry have been provided, it is possible to proceed further and try to correlate the results computed with the CtFD analysis to the Reynolds number Re and to other dimensionless quantities based on the meander flow geometry. In this paragraph, the pressure drop computed with the hydraulic analysis (§ 4.2.4) will be correlated in form of the dimensionless quantity pressure drop coefficient ζ. Then, correlations will be derived for the turbulent and for the laminar flow.

The pressure drop coefficient is slightly modified with respect to Eq. 5.8 as follows:

$$\zeta = 2 \cdot \Delta p_{\text{layer}} \cdot \rho_{\text{He}}(p_{\text{He}}, T_{\text{He}}) \cdot \left(\frac{\dot{m}}{A_{\text{He}}}\right)^{-2}, \tag{5.9}$$

where Δp_{layer} is the pressure drop occurring over one layer of the computational domain, or the half of the total pressure drop occurring between the inlet and the outlet of the computational domain. Indeed, the assumption of partially periodic inlet and outlet boundary conditions leads the pressure drop to be evenly split over the two layers.

For each meander flow geometry listed in Tab. 4.3, the pressure drop coefficients have been calculated for each Reynolds number Re discrete positions at which the helium thermal-fluid mechanics has been investigated (as explained in Tab. 4.4). In case the Reynolds number is $Re > 2000$, the pressure drop coefficient is calculated using the pressure drop computed with the turbulent modelling; In case the Reynolds number is $Re < 1000$, the pressure drop coefficient is calculated using the pressure drop computed with the laminar modelling.

At this point, for each meander flow geometry listed in Tab. 4.3 there are two sets of Reynolds numbers Re and pressure drop coefficient ζ: one set for the turbulent regime, i.e. for $Re = 2000$ - 30000, and one set for the laminar regime, i.e. for $Re = 250$ - 1000.

According to the Buckingham's theorem [Buc14], it is possible to find a general correlation for both the turbulent and the laminar regime as:

$$\zeta = g\left(Re, x_{\mathrm{j}}\right), \tag{5.10}$$

where x_{j}, j = 1,2…n are dimensionless ratios of meander flow geometrical parameters. Equation 5.10 allows the calculation of the pressure drop coefficient from the Reynolds number and some meander flow geometry parameters.

The function g has the form of a power law, or:

$$\zeta = Re^{a} \cdot \prod_{\mathrm{j}} x_{\mathrm{j}}^{b_{\mathrm{j}}}, \tag{5.11}$$

where the exponents b_{j} are determined with a unique multivariate regression analysis.

5.3.1 Correlation for ζ in the turbulent regime

In the first section of this Chapter, the analysis of the turbulent flow between the fins of the heat exchanger has provided important guidelines for the definition of the characteristic helium cross section A_{He} and of the hydraulic diameter d_{h}. Nevertheless, for an exhaustive characterization of the pressure drop in the meander flow geometry it is also necessary to investigate:

- the interaction of the flow with the central bar, and
- the behaviour of the flow in the cut-off region.

Interaction of the helium flow with the central bar

The flow over an obstacle experiences a pressure drop related to the skin friction and to the flow deformation due to the shape of the object itself [KSA87, p. **6**•5]. The relative contribution of these phenomena is not constant and depends on the flow conditions, i.e. on the Reynolds number. As the Reynolds number increases, the contribution to the pressure drop due to the shape of the obstacle increases and largely overcomes the contribution due to the friction [KSA87, p. **6**•7].

The interaction of the helium stream with the central bar is an example of flow over an obstacle. Under the flow conditions investigated in the present work, the contribution due to the shape of the obstacle is expected to be dominant with respect to the losses due to the friction [KSA87, p. 6•7]. The interaction with the central bar determines the flection of the flow with a local fluid acceleration at some position around the central bar itself. Red regions in Fig 5.2 and Fig. 5.3 indicate areas with higher helium velocities. The meander flow geometrical parameters that influence the interaction of the helium flow with the central bar are the central bar diameter d_i itself, the outer diameter d_o and the cut-off c_o. These quantities have been arranged in two dimensionless ratios, namely d_i/d_o and c_o/d_o.

Helium flow in the cut-off region

The topology of the cut-off region is the same as that of a particular bended duct, or elbow, named Π-elbow [Ide86, p. 307]. From this point of view, four meander flow geometrical quantities are relevant for describing the flow in the cut-off: the fin distance t, the fin thickness s, the helium cross section A_{He} and the cross section of the channel in the cut-off region A_{co}. The cross section of the channel in the cut-off region is shown in Fig 5.1 a) and can be calculated with the following formula:

$$A_{co} = \left(\pi \cdot d_o^2 \cdot \frac{\alpha}{720}\right) - \left(\frac{d_o}{2} - c_o\right) \cdot \left(\frac{d_o}{2} \sin(\alpha)\right), \tag{5.12}$$

where

$$\alpha = \operatorname{acos}\left(\left(\frac{d_o}{2} - c_o\right) \Big/ \frac{d_o}{2}\right). \tag{5.13}$$

In more detail, it is not each value of the four quantities per se to determine the behaviour of the flow in the cut-off region, but rather the two ratios s/t and A_{co}/A_{He} (they are the equivalent version for the meander flow geometry of those discussed in [Ide86, p. 307] for the Π-elbow). The turbulent flow has been therefore analysed for several values of both these ratios. Results for increasing s/t values in the range 0.10 - 2.0 and constant A_{co}/A_{He} ratio are shown in Fig. 5.8. Results for increasing A_{co}/A_{He} values in the range 0.89 – 6 and constant s/t ratios are shown in Fig. 5.9. As it can be clearly noted, the characteristic of the helium flow in terms of recirculation regions, their dimension and displacement depend upon the ratios s/t and A_{co}/A_{He}. Therefore, they likely influence the total pressure drop occurring across the computational domain.

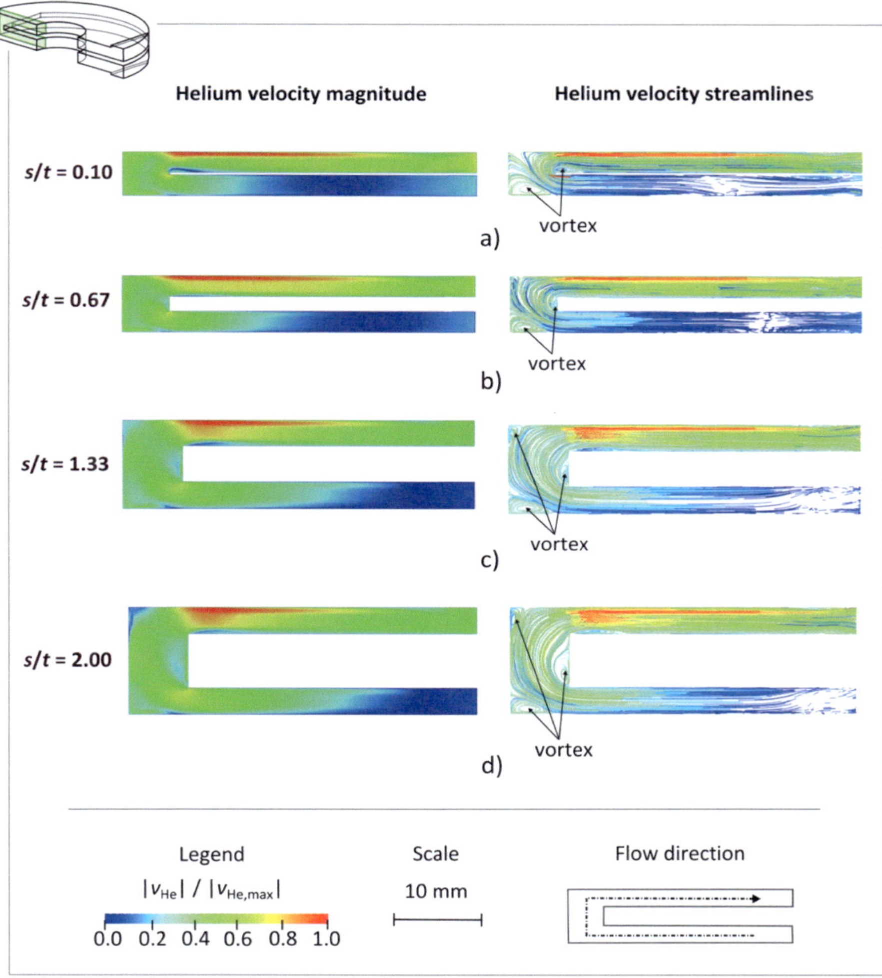

Fig. 5.8: Turbulent helium flow in the cut-off region (Re > 20000), increasing s/t and constant A_{co}/A_{He}:

a) Macro group III, meander flow geometry #17, Tab 4.3
d_o = 140 mm, d_i = 45 mm, t = 3 mm, s = 0.3 mm, c_o = 7 mm;

b) Macro group I, meander flow geometry #4, Tab 4.3
d_o = 140 mm, d_i = 45 mm, t = 3 mm, s = 2 mm, c_o = 7 mm;

c) Macro group III, meander flow geometry #18, Tab 4.3
d_o = 140 mm, d_i = 45 mm, t = 3 mm, s = 4 mm, c_o = 7 mm;

d) Macro group III, meander flow geometry #19, Tab 4.3
d_o = 140 mm, d_i = 45 mm, t = 3 mm, s = 6 mm, c_o = 7 mm.

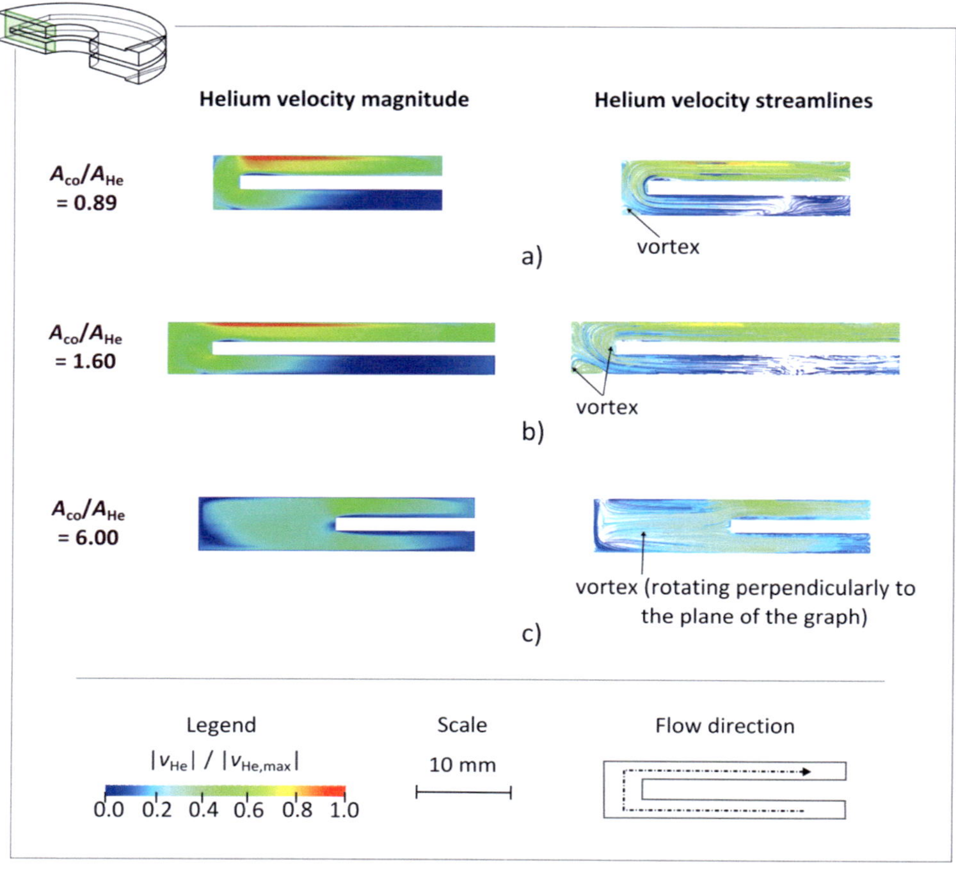

Fig. 5.9: Turbulent helium flow in the cut-off region ($Re > 20000$), increasing A_{co}/A_{He} and constant s/t:

a) Macro group IV, meander flow geometry #20, Tab 4.3
$d_o = 120$ mm, $d_i = 55$ mm, $t = 3$ mm, $s = 2$ mm, $c_o = 4$ mm;

b) Macro group I, meander flow geometry #6, Tab 4.3
$d_o = 160$ mm, $d_i = 65$ mm, $t = 3$ mm, $s = 2$ mm, $c_o = 7$ mm;

c) Macro group I, meander flow geometry #12, Tab 4.3
$d_o = 200$ mm, $d_i = 120$ mm, $t = 3$ mm, $s = 2$ mm, $c_o = 20$ mm.

Correlation for the pressure drop coefficient ζ in the turbulent regime

According to the previous considerations, the correlation for the pressure drop coefficient in turbulent regime contains four dimensionless ratios x_j, namely: d_i/d_o, c_o/d_o, s/t and A_{co}/A_{He}.

The correlation has the mathematical form as Eq. 5.11 and the exponents have been determined with a multivariate regression analysis. The resulting equation is as follows:

$$\zeta = 13.5 \cdot (Re)^{-0.12} \cdot \left(\frac{d_i}{d_o}\right)^{0.16} \cdot \left(\frac{c_o}{d_o}\right)^{-0.07} \cdot \left(\frac{s}{t}\right)^{-0.23} \cdot \left(\frac{A_{co}}{A_{He}}\right)^{-0.15} . \tag{5.14}$$

In Fig. 5.10 the ζ values calculated with the correlation in Eq. 5.14 are plotted against those computed in the CtFD analysis: the maximum deviation is 28%, whereas the average deviation is about 9%. To measure how well future outcomes are likely to be predicted by the correlation, the classical definition of the coefficient of determination R2 has been used (R2 = 1 - residual sum of squared errors for the fitted model/ total sum of squares). In this case R2 is 70%.

The relatively low value of R2 depends on the spread of the points around the bisector, which does not show a random pattern: this could mean that either the correlation does not catch entirely the physics characterizing the problem, or that its mathematical structure might not be the most appropriate. Nevertheless, to the best of the author's knowledge this is the first attempt to correlate the pressure drop coefficient to meander flow geometries whose geometrical parameters span over such broad ranges (see Tab. 4.3).

The ranges of the parameters from which the correlation has been derived, and therefore inside which it is applicable, are summarized in Tab. 5.1.

Tab. 5.1: Range of applicability for the dimensionless parameters in Eq. 5.14.

Re	2000 - 30000
d_i/d_o	0.32 – 0.75
c_o/d_o	0.029 – 0.11
s/t	0.10 – 2.0
A_{co}/A_{He}	0.54 – 6.81

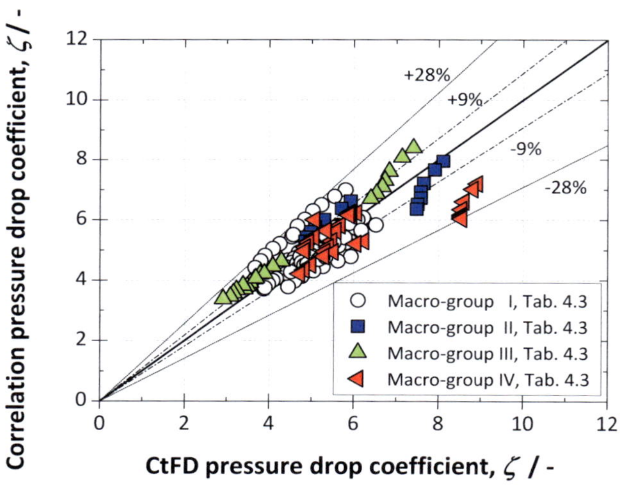

Fig. 5.10: Pressure drop coefficient derived with the correlation in Eq. 5.14 vs. computed pressure drop coefficient.

5.3.2 Correlation ζ in the laminar regime

The characterization of the pressure drop in the laminar regime and the derivation of a correlation for the pressure drop coefficient follow the same approach as for the turbulent regime. The computed helium flow field in several meander flow geometries is analysed. The aim is to identify how the meander flow geometrical parameters influence the helium flow both between the fins of the heat exchanger and in the cut-off region.

Helium flow between the fins

The laminar helium flow is analysed in meander flow geometries having the ratio $(A_0 / (d_o - d_i) \cdot t)$ both smaller and larger than one. Results for the first case are shown in Fig. 5.11, whereas results for the second case are shown in Fig. 5.12. For the sake of generality, the helium velocity $\boldsymbol{v}_{He}$ has been normalized to its maximum value.

In Fig. 5.11 it can be seen that, as for the corresponding turbulent case, a main flow channel exists in the laminar regime as well. This region is characterized by a high helium velocity, which decreases at the upper side of the section. Unlike the turbulent case, the analysis of the helium flow streamlines reveals vortices that are smaller and less persistent. Also for small values of the ratio (A_0 / (d_o - d_i)·t), the recirculation which originates from the detachment of the stream from the external wall is limited to the inlet region (on the right side). The laminar helium flow has therefore the tendency to expand as it flows between the fins.

The interaction with the central bar is less intense than for the turbulent regime. Indeed there is almost no detachment of the helium stream on the left side of the central bar.

Also in meander flow geometries having the ratio (A_0 / (d_o - d_i)·t) larger than one the helium expands as it flows between the fins. Indeed, basically no recirculation region originates after the inlet. Nevertheless, this effect is not as relevant as in the case $A_0 < (d_o - d_i)\cdot t$.

The interaction with the central bar is less intense than for the turbulent regime, though, in this case as well, vortices originate close to the cut-off region for large values of the ratio (A_0 / (d_o - d_i)·t).

To quantify the behaviour of the laminar helium flow between the fins, the dimensionless parameter β has been introduced. The parameter β is defined as:

$$\beta = \frac{A_0}{(d_o - d_i)\cdot t}, \tag{5.15}$$

if the helium flow cross section A_{He} is equal to A_0, or if the ratio (A_0 / (d_o - d_i)·t) is smaller than one. On the contrary, the dimensionless parameter is set to $\beta = 1$ if the helium flow cross section A_{He} is set equal to (d_o - d_i)·t, or if the ratio (A_0 / (d_o - d_i)·t) is larger than one.

As for the turbulent case, the interaction with the central bar is described with the dimensionless quantities d_i/d_o and c_o/d_o.

Helium flow in the cut-off region

As mentioned above, the cut-off region of the meander flow geometry is similar to a particular type of elbow named Π-elbow [Ide86, p. 307]. For the turbulent regime, the flow in this region depends upon the ratios s/t and A_{co}/A_{He}. For the laminar regime, the influence of the cut-off is much weaker. The dimensionless parameters s/t and A_{co}/A_{He} are not relevant for the characterization of the pressure drop and therefore for the characterization of the pressure drop coefficient ζ.

Correlation for the pressure drop coefficient ζ in the laminar regime

The correlation for the pressure drop coefficient in laminar regime contains the dimensionless ratios d_i/d_o, c_o/d_o, and β. No dimensionless ratios have been considered involving the cut-off region, because of their irrelevance in this case. On the contrary, it has been found that the pressure drop coefficient ζ is somehow influenced by the ratio s/d_i.

The correlation, derived in the same mathematical form as Eq. 5.11, is:

$$\zeta = 143 \cdot (Re)^{-0.52} \cdot \left(\frac{d_i}{d_o}\right)^{0.49} \cdot (\beta)^{-0.27} \cdot \left(\frac{c_o}{d_o}\right)^{-0.17} \cdot \left(\frac{s}{d_i}\right)^{-0.07} . \tag{5.16}$$

In Fig. 5.13 the ζ values calculated with the correlation in Eq. 5.16 are plotted against those computed in the CtFD analysis: the maximum deviation between the computed data and the correlation is 26%, whereas the average deviation is about 7%. To measure how well future outcomes are likely to be predicted by the correlation, the classical definition of the coefficient of determination R2 has been used. In this case, R2 ~ 82%.

The ranges of the parameters from which the correlation has been derived, and therefore inside which it is applicable, are summarized in Tab. 5.2.

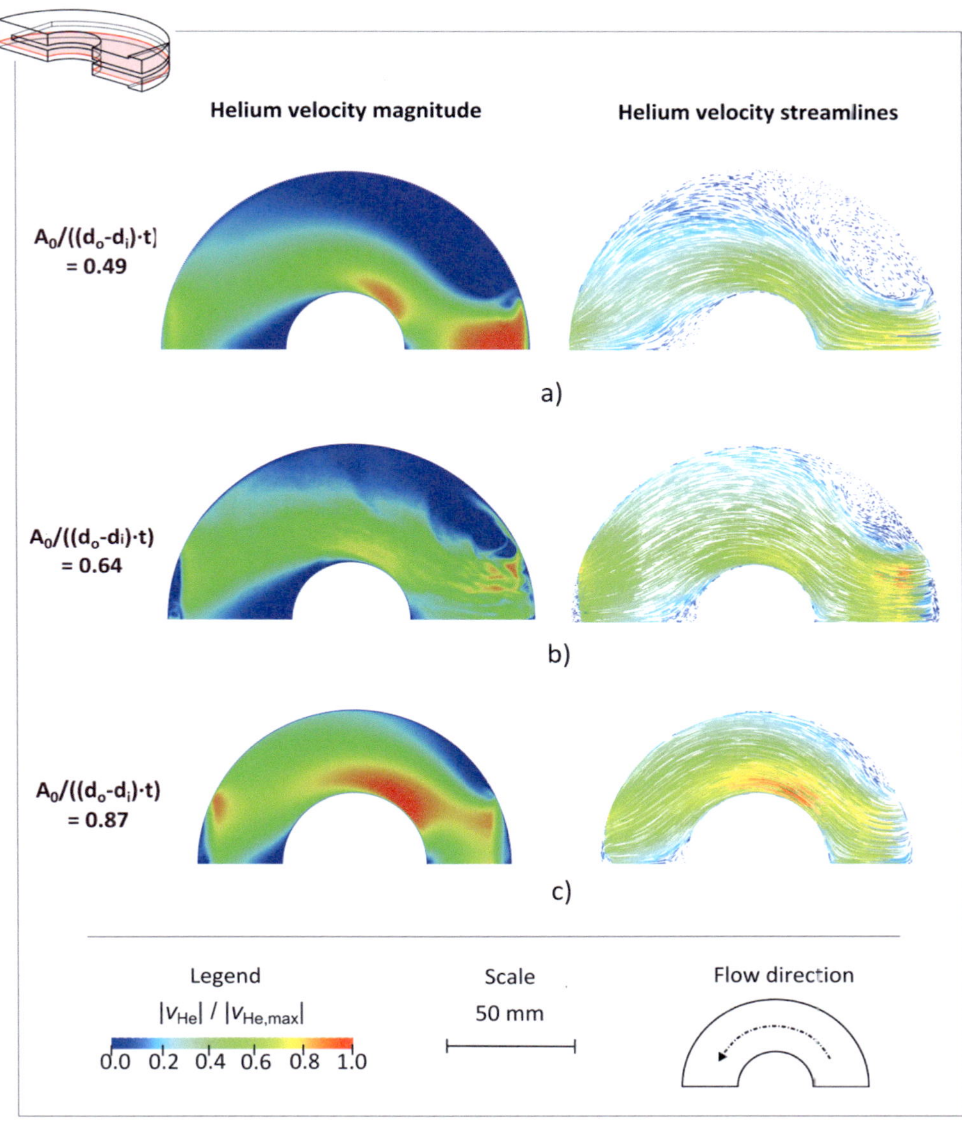

Fig. 5.11: Laminar helium flow between the fins ($Re < 500$), $A_0 < (d_o - d_i) \cdot t$:

a) Macro group IV, meander flow geometry #22, Tab 4.3
$d_o = 140$ mm, $d_i = 45$ mm, $t = 5$ mm, $s = 2$ mm, $c_o = 4$ mm;

b) Macro group II, meander flow geometry #13, Tab 4.3
$d_o = 140$ mm, $d_i = 45$ mm, $t = 5$ mm, $s = 2$ mm, $c_o = 7$ mm;

c) Macro group I, meander flow geometry #2, Tab 4.3
$d_o = 120$ mm, $d_i = 55$ mm, $t = 3$ mm, $s = 2$ mm, $c_o = 7$ mm.

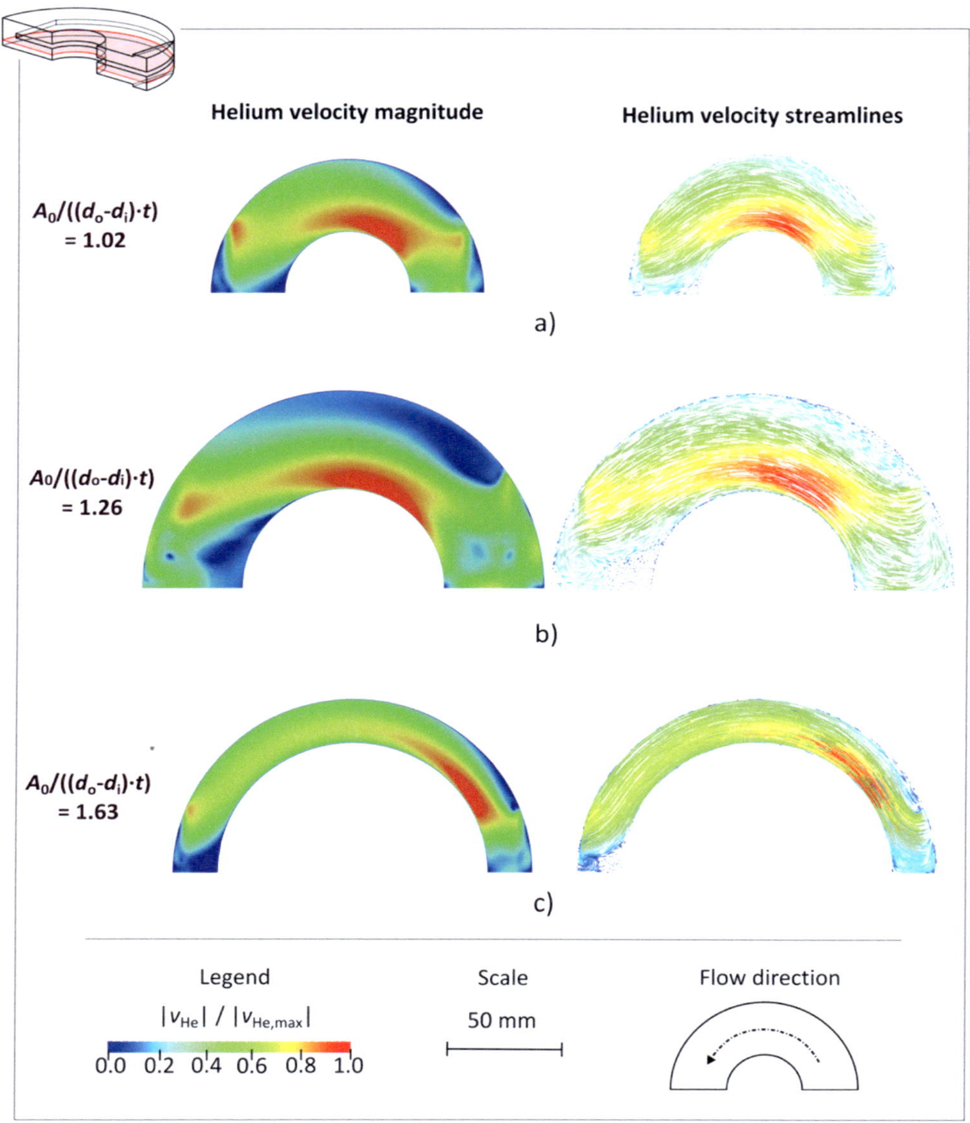

Fig. 5.12: Laminar helium flow between the fins ($Re < 500$), $A_0 > (d_o - d_i)\cdot t$:

a) Macro group IV, meander flow geometry #21, Tab 4.3
$d_o = 120$ mm, $d_i = 55$ mm, $t = 3$ mm, $s = 2$ mm, $c_o = 10$ mm;

b) Macro group II, meander flow geometry #16, Tab 4.3
$d_o = 180$ mm, $d_i = 90$ mm, $t = 8$ mm, $s = 2$ mm, $c_o = 20$ mm;

c) Macro group I, meander flow geometry #9, Tab 4.3
$d_o = 160$ mm, $d_i = 120$ mm, $t = 3$ mm, $s = 2$ mm, $c_o = 7$ mm.

Tab. 5.2: Range of applicability for the dimensionless parameters in Eq. 5.16.

Re	250 - 1000
d_i/d_o	0.32 – 0.75
β	0.49 – 1
c_o/d_o	0.029 – 0.11
s/d_i	0.0067 – 0.13

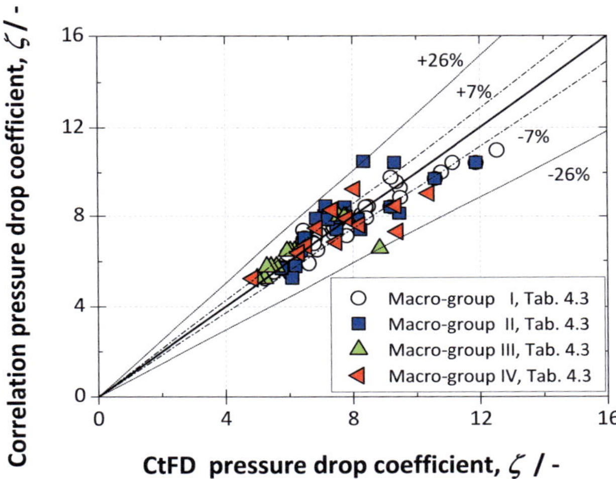

Fig. 5.13: Pressure drop coefficient derived with the correlation in Eq. 5.16 vs. computed pressure drop coefficient.

5.4 Correlation for the Nusselt number, *Nu*

The last section of this Chapter is about the derivation of correlations for the Nusselt number Nu for both the laminar and turbulent regime.

As explained in the previous Chapter (§ 4.2), the Nusselt number is the ratio of the convective to the conductive heat transfer occurring in direction perpendicular to the fluid and is defined as:

$$Nu = \frac{h \cdot d_{\mathrm{h}}}{\lambda_{\mathrm{He}}(p_{\mathrm{He}}, T_{\mathrm{He}})}, \tag{5.17}$$

where d_{h} is the hydraulic diameter, λ_{He} is the thermal conductivity of the helium and h is the heat transfer coefficient. The heat transfer coefficient is calculated from the results of the thermal-hydraulic analysis of the helium flow in the meander flow geometry (§ 4.2.4). The thermal-hydraulic analysis uses the helium flow field computed in the hydraulic analysis as an input to evaluate the macroscopic energy transport. A temperature $T_{\mathrm{w}} > T_{\mathrm{He,in}}$ was applied to the wall boundary conditions, as indicated in § 4.2.4. The aim is to calculate the temperature difference ΔT_{He} between the inlet and the outlet of the periodic model, which is defined as:

$$\Delta T_{\mathrm{He}} = T_{\mathrm{He,out}} - T_{\mathrm{He,in}}. \tag{5.18}$$

The heat transfer coefficient is obtained by performing the heat balance over the computational domain:

$$h = \frac{\dot{m} \cdot c_{\mathrm{p,He}} \cdot \Delta T_{\mathrm{He}}}{A_{\mathrm{HX}} \cdot \left(T_{\mathrm{w}} - T_{\mathrm{b,He}}\right)}, \tag{5.19}$$

where A_{HX} is the area of the heated surfaces, or the area of the surfaces on which the temperature T_{w} had been imposed; the term $T_{\mathrm{b,He}}$ is the helium flow bulk temperature, or the mean temperature of the helium averaged on the flow field inside in the computational domain. The averaging of the helium temperature on the flow field is necessary because, as extensively demonstrated by the above-mentioned results, the helium flow in the meander flow geometry is not homogeneous. The helium temperature increases more in the recirculation regions, for instance. Locally displaced hot spots may therefore originate, which do not interact with the helium main channel. The evaluation of the helium flow bulk temperature based on the flow fields provides therefore a better description of the convective cooling effectiveness. Practically, the helium flow bulk temperature has been calculated from the CtFD analysis results as follow:

$$T_{\mathrm{b,He}} = \frac{\sum_{\mathrm{i}} T_{\mathrm{He,i}} \cdot v_{\mathrm{i}} \cdot V_{\mathrm{i}}}{\sum_{\mathrm{i}} v_{\mathrm{i}} \cdot V_{\mathrm{i}}}, \tag{5.20}$$

where $T_{\mathrm{He,i}}$ is the helium temperature, v_{i} the helium velocity and V_{i} the volume of the i-th computational cell in the model.

The derivation of the correlations for the Nusselt number Nu in laminar and turbulent regime follows an analogous procedure as the derivation of the correlations for the pressure drop coefficient: based on the Buckingham's theorem [Buc14], general correlations for both the turbulent and the laminar regime are searched in the form:

$$Nu = f(Re, x_{\mathrm{i}}), \tag{5.21}$$

where x_{i}, i = 1,2…n are dimensionless ratios of meander flow geometrical parameters.

In this case as well, the function f has the form of power law:

$$Nu = Re^{a} \cdot \prod_{\mathrm{i}} x_{\mathrm{i}}^{b_{\mathrm{i}}}, \tag{5.22}$$

where the exponents b_{i} are determined with a unique multivariate regression analysis.

5.4.1 Correlation for *Nu* in the turbulent regime

In order to quantify the effect of the meander geometry on the temperature distribution inside the helium flow, the results of the thermal-hydraulic analysis have been studied with particular emphasis on the region between the fins.

As it is already known from § 5.1.1, the turbulent helium flow between the fins is strongly influenced by the ratio $(A_0 / (d_o - d_i) \cdot t)$. Consequently, also the temperature distribution inside the helium flow is influenced by this ratio. Two case studies are proposed in Fig. 5.14 and Fig. 5.15, respectively for $(A_0 / (d_o - d_i) \cdot t) < 1$ and $(A_0 / (d_o - d_i) \cdot t) > 1$. The temperature distribution has been plotted on each mid-plane cutting the two layers of the computational domain. To generalize the handling of the results, the temperature has been normalized with the introduction of the parameter $(T_{He} - T_{He,in}) / (T_w - T_{He,in})$. Furthermore, also the helium velocity streamlines have been plotted; the purpose is to highlight the dependence of the temperature distribution with respect to the helium flow. In both cases, the Reynolds number is $Re > 20000$. As it can be seen, in Fig 5.14, the helium temperature in vortex regions almost increases to the wall temperature T_w. Since the temperature difference with respect to the wall is very small, the heat flux removed in the recirculation regions is limited. Similarly limited is also the conductive heat transfer between the recirculation regions and the main flow channel. This is due to the low helium thermal conductivity λ_{He}. A quite different behaviour is observed in Fig. 5.15, for the case where $(A_0 / (d_o - d_i) \cdot t) > 1$. Indeed, the helium flow heats up more homogeneously along the computational domain.

The analogy between the helium flow and the helium temperature distribution between the fins suggests that the Nusselt number Nu might be correlated to some of the dimensionless ratios influencing the pressure drop coefficient ζ (Eq. 5.14). This has been proven to be true for the dimensionless quantities d_i/d_o and c_o/d_o.

Unlike the pressure drop coefficient, the Nusselt number Nu in turbulent regime does not show any relevant dependence on the dimensionless quantities s/t and A_{He}/A_{co}.

On the other hand, the Nusselt number varies with the dimensionless ratios t/d_i and A_{He}/A_{HX}. The quantity A_{HX} is the heat transfer surface between three neighbouring fins and can be calculated as:

$$A_{HX} = \left(4 \cdot \left(\frac{d_o^2}{2} - \frac{d_i^2}{2}\right) + 2 \cdot (d_i \cdot t)\right) \cdot \pi - 3 \cdot A_{co}. \tag{5.23}$$

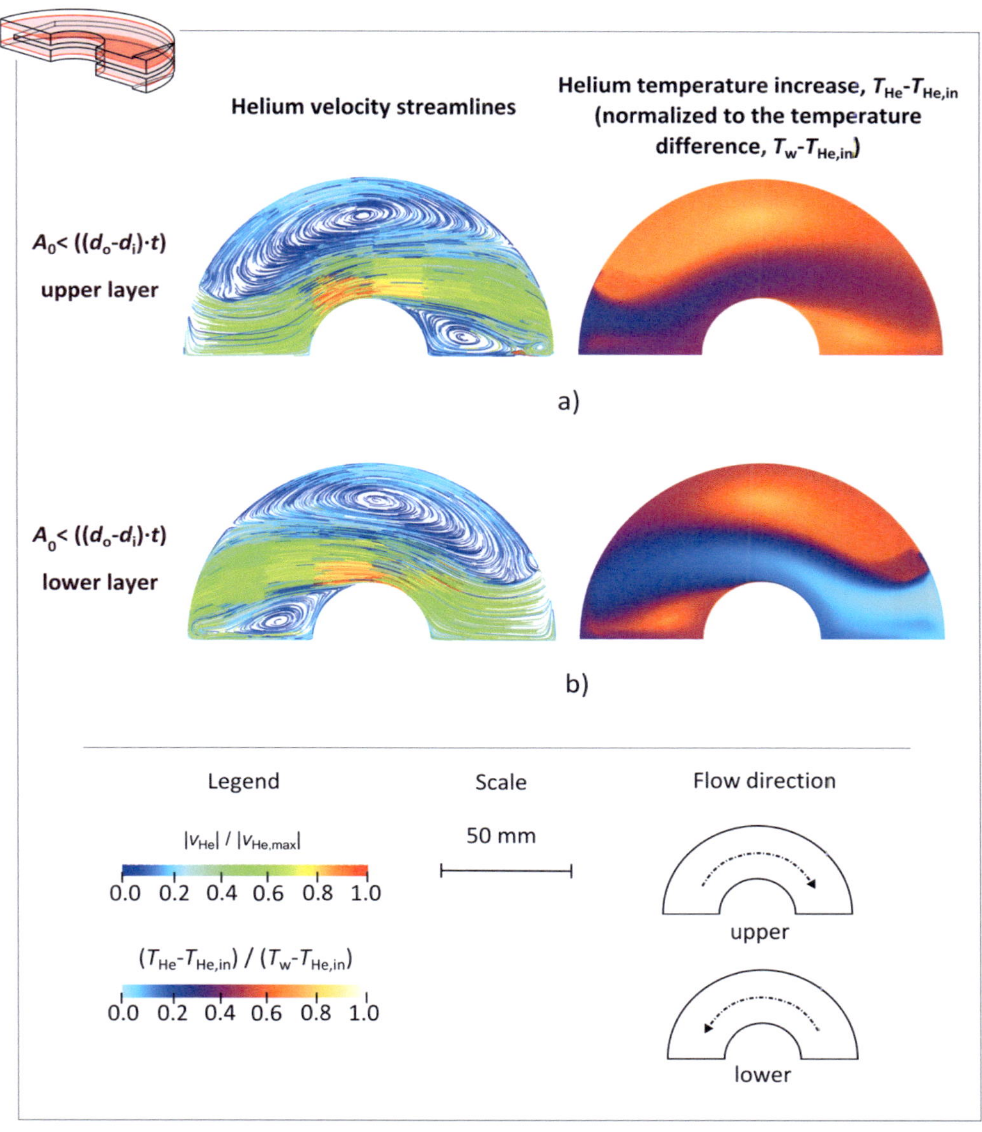

Fig. 5.14: Turbulent helium flow and helium temperature distribution between the fins (Re > 20000), $A_0 < (d_o - d_i) \cdot t$;

Macro group II, meander flow geometry #13, Tab 4.3
d_o = 140 mm, d_i =·45 mm, t = 5 mm, s = 2 mm, c_o = 7 mm;

a) Lower layer;

b) Upper layer.

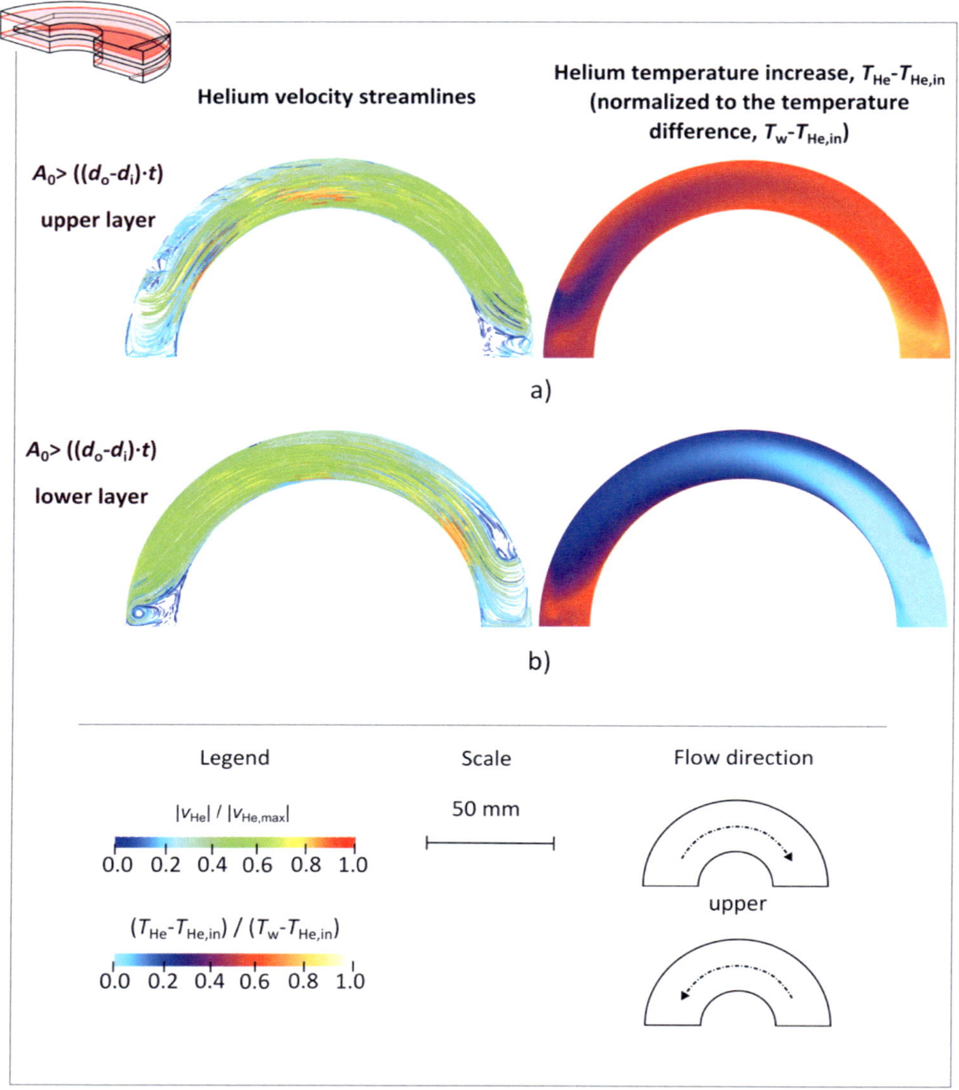

Fig. 5.15: Turbulent helium flow and helium temperature distribution between the fins (Re > 20000), $A_0 > (d_o - d_i)\cdot t$;

Macro group I, meander flow geometry #9, Tab 4.3
d_o = 160 mm, d_i =·120 mm, t = 3 mm, s = 2 mm, c_o = 7 mm;

a) Lower layer;

b) Upper layer.

Correlation for the Nusselt number Nu in the turbulent regime

The correlation for the Nusselt number in turbulent regime is derived from the Reynolds number *Re* and the dimensionless quantities d_i/d_o, c_o/d_o, t/d_i and A_{He}/A_{HX}. The form of the correlation is a power law where the exponents are calculated with a multivariate regression analysis:

$$Nu = 0.84 \cdot (Re)^{0.73} \cdot \left(\frac{A_{He}}{A_{HX}}\right)^{0.15} \cdot \left(\frac{d_i}{d_o}\right)^{0.72} \cdot \left(\frac{t}{d_i}\right)^{0.24} \cdot \left(\frac{c_o}{d_o}\right)^{0.12}. \tag{5.24}$$

The values of *Nu* calculated with the correlation are plotted against those computed with the CtFD analysis: the maximum deviation is 20%, whereas the average deviation is about 5%. The parameter R2 is 98%. In Fig. 5.16 the *Nu* values calculated with the correlation in Eq. 5.24 are plotted against those computed in the CtFD analysis.

The ranges of the parameters from which the correlation has been derived, and therefore inside which it is applicable, are summarized in Tab. 5.3.

It is worth mentioning that, although the term A_{He}/A_{HX} is somehow related to the heat transfer surface of the computational domain, the ratio does not lead to any loss of generality. Indeed, it characterizes the ratio of the helium flow cross section to a multiple of the heat transfer surface between three fins of a given meander flow geometry. A different choice of A_{HX}, i.e. the heat transfer surface between two fins instead ($A_{HX}/2$) or between seven fins ($3 \cdot A_{HX}$), would only change the coefficient of the Eq. 5.24. The exponent of the dimensionless ratio remains the same.

Tab. 5.3: Range of applicability for the dimensionless parameters in Eq. 5.24.

Parameter	Range
Re	2000 - 30000
d_i/d_o	0.32 – 0.75
c_o/d_o	0.029 – 0.11
t/d_i	0.025 – 0.18
A_{He}/A_{HX}	0.0029 – 0.0096

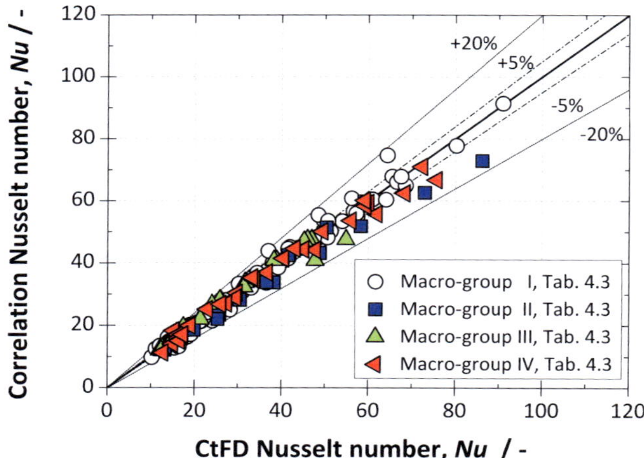

Fig. 5.16: Nusselt number derived with the correlation in Eq. 5.24 vs. computed Nusselt number.

5.4.2 Correlation for *Nu* in the laminar regime

For laminar flows in simple geometries, the Nusselt number *Nu* is generally expressed by a pure number. It typically depends only on the specific geometry and not on the flow conditions, or on the Reynolds number *Re* [Bej84, p. 78]. Nevertheless this is not the case for the meander flow geometry, where the heat transfer is weakly dependent on the Reynolds number. This is of no surprise, considering the complexity of the flow field.

Besides the Reynolds number, other dimensionless quantities related to the meander flow geometry are expected to influence the Nusselt number. As for the turbulent regime case, the temperature distribution between the fins is therefore analysed for two case studies, respectively with $(A_0 / (d_o - d_i) \cdot t) < 1$ and $(A_0 / (d_o - d_i) \cdot t) > 1$. Results are shown in Fig. 5.17 and Fig. 5.18. To generalize the handling of the results, the temperature has been normalized with the introduction of the parameter $(T_{He} - T_{He,in}) / (T_w - T_{He,in})$. Also the helium velocity streamlines have been plotted with the purpose of comparison between the temperature distribution and the helium flow. A relevant difference can be immediately recognized with respect to the turbulent equivalent case in Fig. 5.14 and Fig. 5.15: in the laminar regime, despite the $(A_0 / (d_o - d_i) \cdot t)$, the helium flow heats up soon after the inflow between the fins. This is related to the different behaviour of the flow which, as stated above, has the tendency to expand in the fin region. The dimensionless parameter β likely has an influence on the Nusselt number as well.

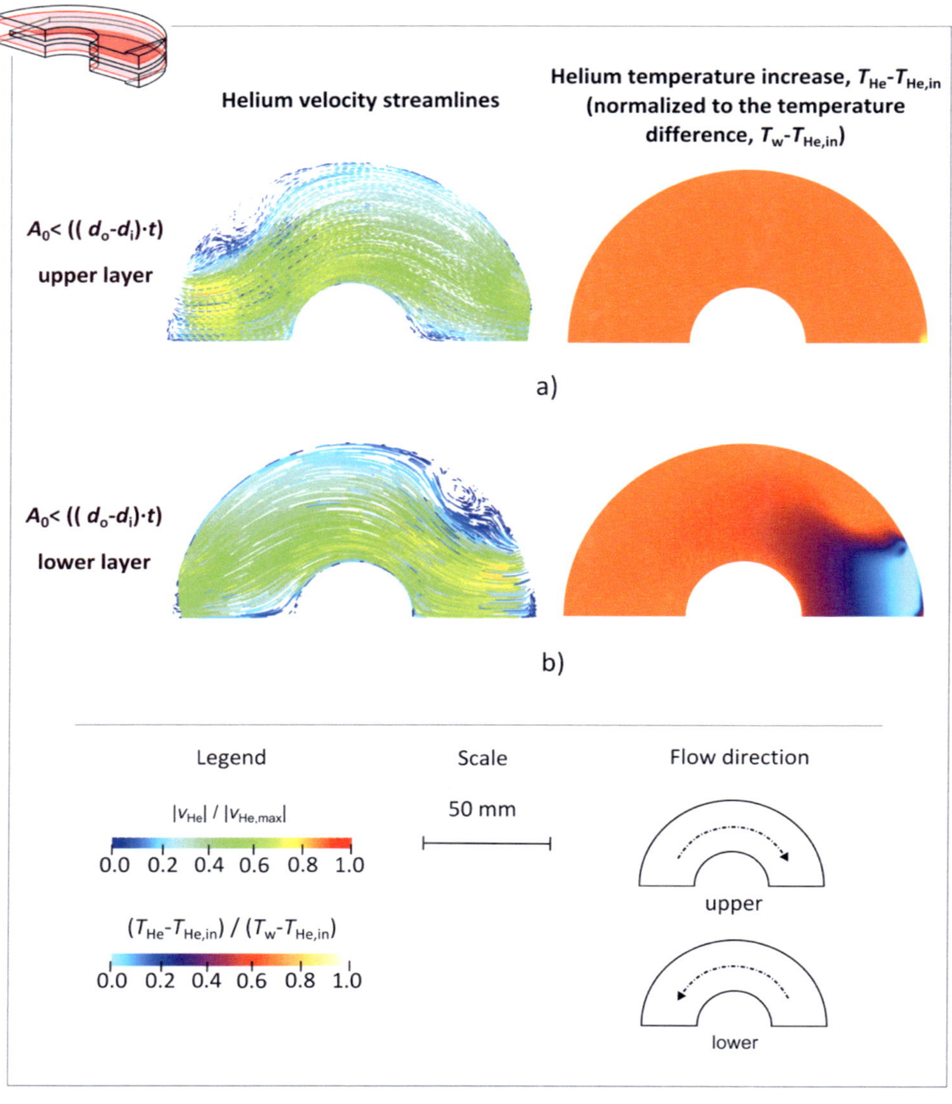

Fig. 5.17: Laminar helium flow and helium temperature distribution between the fins (Re < 500), $A_0 < (d_o - d_i)\cdot t$;

Macro group II, meander flow geometry #13, Tab 4.3
d_o = 140 mm, d_i =·45 mm, t = 5 mm, s = 2 mm, c_o = 7 mm;

a) Lower layer;

b) Upper layer.

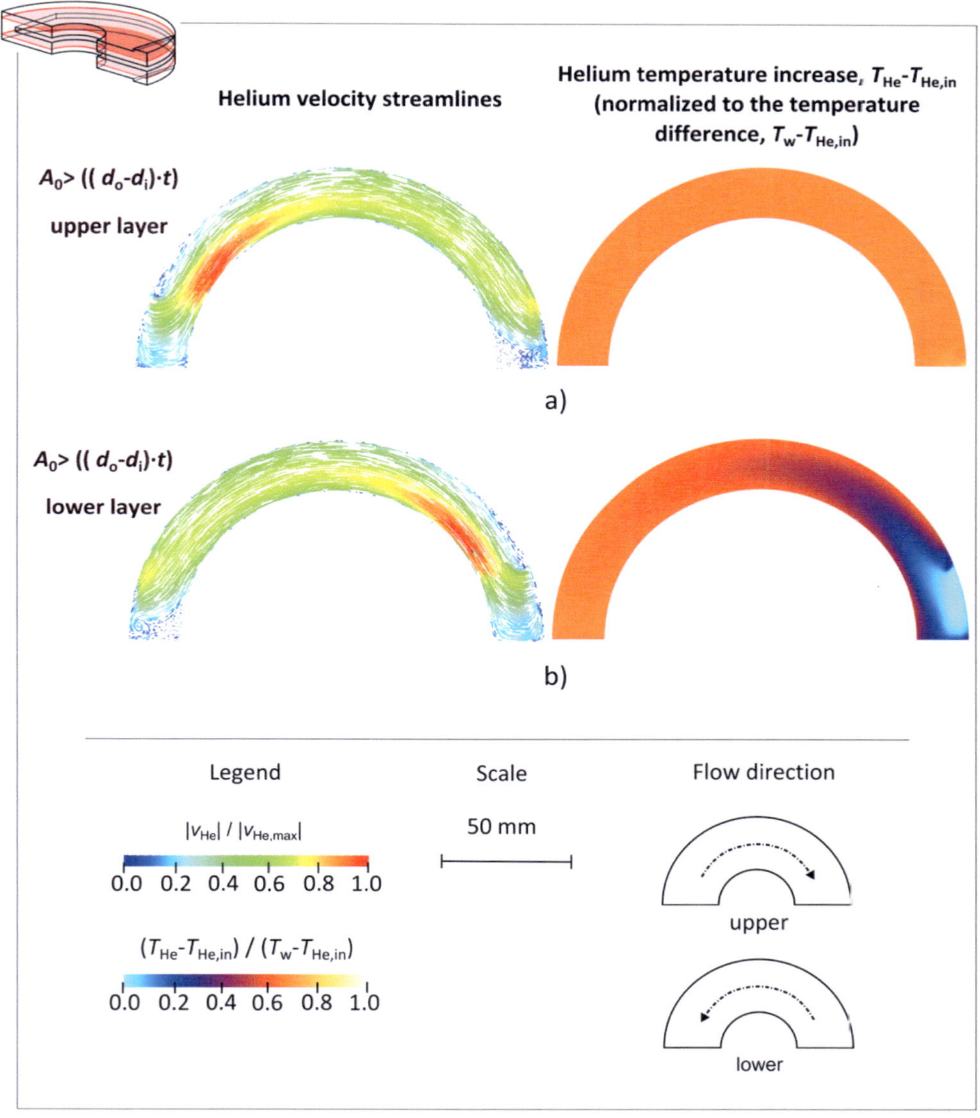

Fig. 5.18: Laminar helium flow and helium temperature distribution between the fins (Re < 500), $A_0 > (d_o - d_i)\cdot t$;

Macro group I, meander flow geometry #9, Tab 4.3
d_o = 160 mm, d_i = 120 mm, t = 3 mm, s = 2 mm, c_o = 7 mm;

a) Lower layer;

b) Upper layer.

With the definition of the characteristic geometrical quantities helium cross section A_{He} and hydraulic diameter d_{h} it has been explained that, to a certain extent, the helium flow inside the meander flow geometry is similar to the flow in a rectangular duct. In this kind of duct, the Nusselt number in laminar regime depends on the "slenderness" of the channel [KSA87, p. **3•**47]. In the meander flow geometry, the slenderness of the helium channel is given by the parameter t/a. Actually, the computed Nusselt numbers show a certain dependence on t/a.

Similarly to the turbulent regime case, the Nusselt number in the laminar regime is influenced by the ratio t/d_{i} as well. Also the dimensionless ratio s/d_{o} plays a role.

Correlation for the Nusselt number *Nu* in laminar regime

The correlation for the Nusselt number in turbulent regime is derived from the Reynolds number *Re* and the dimensionless quantities β, t/d_{i}, t/a and s/d_{o}. The form of the correlation is a power law where the exponents are calculated with a multivariate regression analysis:

$$Nu = 20 \cdot (Re)^{0.10} \cdot \left(\frac{s}{d_{\mathrm{o}}}\right)^{0.15} \cdot \left(\frac{t}{d_{\mathrm{i}}}\right)^{0.08} \cdot (\beta)^{0.63} \cdot \left(\frac{t}{a}\right)^{0.12} . \tag{5.25}$$

The maximum deviation is 13%, whereas the average deviation is about 4.5%. The parameter R2 is R2 ~ 84%. In Fig. 5.19 the *Nu* values calculated with the correlation in Eq. 5.25 are plotted against those computed in the CtFD analysis.

The ranges of the parameters from which the correlation has been derived, and therefore inside which it is applicable, are summarized in Tab. 5.4.

Tab. 5.4: Range of applicability for the dimensionless parameters in Eq. 5.25.

Re	250 - 1000
β	0.49 – 1
t/d_{i}	0.025 – 0.178
s/d_{o}	0.0021 – 0.043
t/a	0.0033 – 0.13

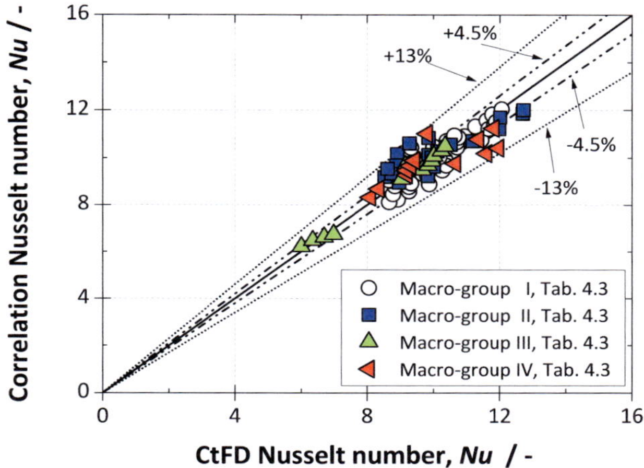

Fig. 5.19: Nusselt number derived with the correlation in Eq. 5.25 vs. computed Nusselt number.

5.5 Summary

In this Chapter, the results of the CtFD periodic modelling have been analysed and the following goals have been achieved:

- formulae to calculate the geometrical quantities characterizing the meander flow geometry, namely the helium flow cross section A_{He} and the hydraulic diameter d_h, have been provided on the basis of the helium turbulent flow analysis,
- the flow regime in the meander flow geometry has been characterized depending on the Reynolds number Re. If the Reynolds number Re = 2000 – 30000, the regime is turbulent; if the Reynolds number Re = 250 – 1000, the regime is laminar. The range Re = 1000 – 2000 has not been directly investigated for the lack of appropriate models; it will be referred to as transition regime,
- the pressure drop computed with the CtFD analysis have been arranged in terms of pressure drop coefficient ζ. Correlations for the pressure drop coefficient in laminar and turbulent flow have been derived with multivariate regression analysis on the Reynolds number and other dimensionless ratios involving meander flow geometry parameters,
- the heat transfer coefficient h computed with the CtFD analysis have been arranged in terms of Nusselt number Nu. Correlations for the Nusselt number in laminar and turbulent flow have been derived with multivariate regression analysis on the Reynolds number and other dimensionless ratios involving meander flow geometry parameters.

These results provide the reader with an accurate description of the helium thermal-fluid mechanics inside the meander flow geometry and make possible to optimize the design of meander flow heat exchangers.

A compendium of the presented formulae is given in the following.

Characteristic geometrical quantities

Hydraulic diameter, d_{h} (§ 5.1.2):

$$d_{\mathrm{h}} = \frac{2 \cdot A_{\mathrm{He}}}{a + t} \sim 2 \cdot t.$$

Helium flow cross section, A_{He} (§ 5.1.1):

$$A_{\mathrm{He}} = \min(A_0, (d_{\mathrm{o}} - d_{\mathrm{i}}) \cdot t),$$

$$A_0 = 2 \cdot \left[\left(\frac{d_{\mathrm{o}}}{2}\right)^2 - \left(\frac{d_{\mathrm{o}}}{2} - c_{\mathrm{o}}\right)^2\right]^{0.5} \cdot t.$$

Width of the helium channel cross section, a (§ 5.1.2):

$$a = (d_{\mathrm{o}} - d_{\mathrm{i}}) \text{ if } A_{\mathrm{He}} = (d_{\mathrm{o}} - d_{\mathrm{i}}) \cdot t,$$

$$a = \left[\left(\frac{d_{\mathrm{o}}}{2}\right)^2 - \left(\frac{d_{\mathrm{o}}}{2} - c_{\mathrm{o}}\right)^2\right]^{0.5} \cdot t \quad \text{if } A_{\mathrm{He}} = A_0.$$

Further geometrical quantities of the meander flow geometry

Expansion rate, β (§ 5.3.2):

$$\beta = 1 \quad \text{if } A_{\mathrm{He}} = (d_{\mathrm{o}} - d_{\mathrm{i}}) \cdot t,$$

$$\beta = \frac{A_0}{(d_{\mathrm{o}} - d_{\mathrm{i}}) \cdot t} \quad \text{if } A_{\mathrm{He}} = A_0.$$

Helium cross section in the cut-off region, A_{co} (§ 5.3.1):

$$A_{\mathrm{co}} = \left(\pi \cdot d_{\mathrm{o}}^2 \cdot \frac{\alpha}{720}\right) - \left(\frac{d_{\mathrm{o}}}{2} - c_{\mathrm{o}}\right) \cdot \left(\frac{d_{\mathrm{o}}}{2} \cdot sin(\alpha)\right),$$

$$\alpha = acos\left(\left(\frac{d_{\mathrm{o}}}{2} - c_{\mathrm{o}}\right) \Big/ \frac{d_{\mathrm{o}}}{2}\right).$$

Heat transfer surface, A_{HX} (§ 5.4.1):

$$A_{\mathrm{HX}} = \left(4 \cdot \left(\frac{d_{\mathrm{o}}^2}{2} - \frac{d_{\mathrm{i}}^2}{2}\right) + 2 \cdot (d_{\mathrm{i}} \cdot t)\right) \cdot \pi - 3 \cdot A_{\mathrm{co}}.$$

Correlation for the pressure drop coefficient, ζ

Definition of the pressure drop coefficient, ζ (§ 5.3):

$$\zeta = 2 \cdot \Delta p_{\text{layer}} \cdot \rho_{\text{He}} \cdot \left(\frac{\dot{m}}{A_{\text{He}}}\right)^{-2}.$$

Correlation for the pressure drop coefficient in laminar flow (Re = 250 – 1000, § 5.3.2):

$$\zeta = 143 \cdot (Re)^{-0.52} \cdot \left(\frac{d_{\text{i}}}{d_{\text{o}}}\right)^{0.49} \cdot (\beta)^{-0.27} \cdot \left(\frac{c_{\text{o}}}{d_{\text{o}}}\right)^{-0.17} \cdot \left(\frac{s}{d_{\text{i}}}\right)^{-0.07}.$$

Correlation for the pressure drop coefficient in turbulent flow (Re = 2000 – 30000, § 5.3.1):

$$\zeta = 13.5 \cdot (Re)^{-0.12} \cdot \left(\frac{d_{\text{i}}}{d_{\text{o}}}\right)^{0.16} \cdot \left(\frac{c_o}{d_{\text{o}}}\right)^{-0.07} \cdot \left(\frac{s}{t}\right)^{-0.23} \cdot \left(\frac{A_{\text{co}}}{A_{\text{He}}}\right)^{-0.15}.$$

Correlation for the Nusselt number, *Nu*

Definition of the Nusselt number, Nu (§ 5.4):

$$Nu = \frac{h \cdot d_{\text{h}}}{\lambda_{\text{He}}}.$$

Correlation for the Nusselt number in laminar flow (Re = 250 – 1000, § 5.4.2):

$$Nu = 20 \cdot (Re)^{0.10} \cdot \left(\frac{s}{d_{\text{o}}}\right)^{0.15} \cdot \left(\frac{t}{d_{\text{i}}}\right)^{0.08} \cdot (\beta)^{0.63} \cdot \left(\frac{t}{a}\right)^{0.12}.$$

Correlation for the Nusselt number in turbulent flow (Re = 2000 – 30000, § 5.4.1):

$$Nu = 0.84 \cdot (Re)^{0.73} \cdot \left(\frac{A_{\text{He}}}{A_{\text{HX}}}\right)^{0.15} \cdot \left(\frac{d_{\text{i}}}{d_{\text{o}}}\right)^{0.72} \cdot \left(\frac{t}{d_{\text{i}}}\right)^{0.24} \cdot \left(\frac{c_o}{d_{\text{o}}}\right)^{0.12}.$$

Limit of applicability of the correlations

Limit of applicability of both correlations for the pressure drop coefficient, ζ, and the Nusselt number, Nu, in laminar regime:

Tab. 5.5: Limit of applicability in the laminar regime.

Re	250 - 1000
d_i/d_o	0.32 – 0.75
β	0.49 – 1
c_o/d_o	0.029 – 0.11
s/d_i	0.0067 – 0.13
t/d_i	0.025 – 0.178
s/d_o	0.0021 – 0.043
t/a	0.0033 – 0.13

Limit of applicability of both correlations for the pressure drop coefficient, ζ, and the Nusselt number, Nu, in turbulent regime:

Tab. 5.6: Limit of applicability in the turbulent regime.

Re	2000-30000
d_i/d_o	0.32 – 0.75
c_o/d_o	0.029 – 0.11
s/t	0.10 – 2.0
t/d_i	0.025 – 0.18
A_{co}/A_{He}	0.54 – 6.81
A_{He}/A_{HX}	0.0029 – 0.0096

6 Numerical modelling of the HTS module

In this Chapter, the techniques developed for the numerical modelling of the HTS module are presented. The computational results are compared with experimental results. The purpose is to define an effective methodology for the modelling of the HTS module in the design phase.

6.1 Design issues of the HTS module

As it has been introduced in the Chapter III, the HTS module operates in normal conditions between the temperature $T_{\mathrm{HTS,W}}$ and $T_{\mathrm{HTS,C}}$. An exemplarily overview on the operation of a HTS module is shown in Fig. 6.1.

Since it is not actively cooled, the cooling power P_{HTS} that has to be provided at the cold end of the HTS current lead depends upon two contributions, as recalled in Eq. 6.1:

$$\begin{aligned} P_{\mathrm{HTS}} &= \left(\dot{Q}_{\mathrm{HTS,cond}} + \dot{Q}_{\mathrm{HTS,loss}}\right) \cdot \frac{1}{\eta_{\mathrm{c},1}} \cdot \frac{1}{e_1} \\ &= \left(\frac{A_{\mathrm{HTS}}}{L_{\mathrm{HTS}}} \cdot \int_{T_{\mathrm{HTS,C}}}^{T_{\mathrm{HTS,W}}} \lambda(T) \cdot dT + \dot{Q}_{\mathrm{HTS,Loss}}\right) \cdot \frac{1}{\eta_{\mathrm{c},1}} \cdot \frac{1}{e_1}, \end{aligned} \tag{6.1}$$

where $\dot{Q}_{\mathrm{HTS,cond}}$ accounts for the heat conducted along the HTS module. The A_{HTS} and L_{HTS} are the cross section and the length of the HTS module, λ(T) is the equivalent heat conductivity, whereas $\eta_{\mathrm{c},1}$ and e_1 are the Carnot and the refrigerator efficiencies at 4.5 K, respectively. The term and $\dot{Q}_{\mathrm{HTS,loss}}$ accounts for the resistive losses inside the HTS module.

Indeed, although the electrical current transport occurs mainly in the HTS superconductors, connections to the heat exchanger and to the bus bar are made of normal conductor (normally copper). Resistive losses cannot be avoided and have to be properly quantified.

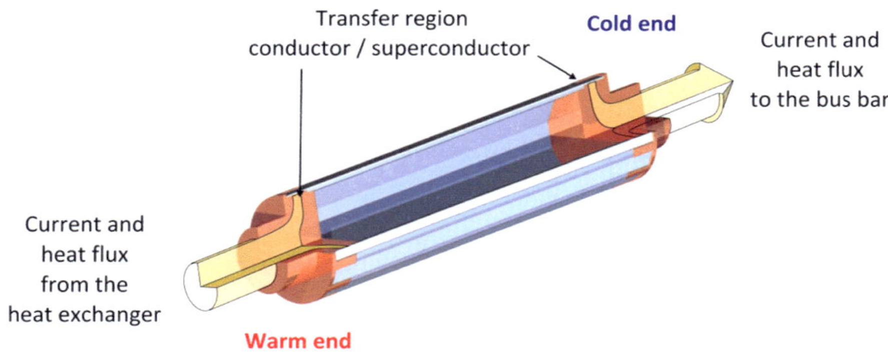

Fig. 6.1: Exemplarily overview of a HTS module in operation between temperature $T_{HTS,W}$ and $T_{HTS,C}$. The arrow entering the HTS module at the warm end and going out at the cold one indicate the direction of the heat flux and, in this case, also of the electrical current (whose direction depends upon the polarity of the current lead).

The existence of connections between normal and superconductors leads to another key issue for the design of a HTS module in normal operation: the transfer length, or the length along which the normal conductor and the superconductor are coupled. The location of these regions is shown in Fig. 6.1. An analogous problem has been handled in [Eki78] and more recently in [SKL07] for the case of current transfer in multi-filamentary superconductors. For simple geometries and considering the transfer from a normal conducting matrix towards a superconducting layer, eventually through a resistive barrier (whose characteristics are known), it is possible to treat the problem analytically [Eki78, SKL07]; by coupling the voltage drop inside the normal conducting matrix and in the superconducting layer to the current conservation law, a homogeneous, second order differential equation for the current in the superconductor is obtained. The analytical solution has the form of an exponential function, from which it is possible to evaluate the length over which a certain electrical current is transferred form the matrix to the superconducting layer. For the case of the transition length HTS superconductors/normal conductors in the HTS module, the above-mentioned analytical approach can only provide a rough estimation. This is due to the simplifying assumptions required for the analytical treatment, which do not allow an exhaustive treatment of the HTS module geometrical arrangement.

Besides the normal operating conditions, the design of the HTS module shall also consider failures or anomalous operations. The most severe accident for a HTS current lead is the Loss Of Flow Accident, LOFA. A LOFA occurs when the active cooling of the resistive heat exchanger fails while the current lead is still in operation. The heat released is deposited inside the structure of the current lead[6] itself and determines a temperature increase that depends on the specific heat capacity c_p of the materials. The heat flux flows counter-parallel to the temperature gradient along the HTS current lead and therefore towards the HTS module. The design of the HTS module has to guarantee a limited temperature increase over a prescribed time interval and prevent the quench of the HTS conductors. Under these conditions, indeed, a thermal runaway could seriously occur with potential severe consequences on the device served by the current lead and the current lead itself.

The design issues described above are gathered in Tab. 6.1 where it is also indicated what kind of analysis has to be applied (i.e. steady-state or time-dependent) to address their analysis.

Tab. 6.1: Design issues of the HTS module and type of analysis required in the design phase.

Type of operation	Design issue of the HTS module	Type of analysis
Normal	Evaluation of P_{HTS} (Eq. 6.1) Evaluation of the transfer length	Steady-state
Fault condition (LOFA)	Temperature increase	Time-dependent

[6] In principle also the casing enclosing the current lead is a bulk in which the thermal energy can be deposited. Nevertheless, the analysis of HTS current leads during a LOFA accident normally considers only the current lead and no extra masses acting as heat capacities. This assumption has at least two important motivations: in the first place, it is intrinsically conservative because a heat capacity smaller than the actual one is considered. In the second place, although conductive heat fluxes towards the casing structure of the current lead are possible, it is worth mentioning that they might occur on a slower time scale than the thermal phenomena inside the current lead.

6.1.1 Computational domains for the modelling of the HTS module

The definition of the computational domain for the numerical analysis takes advantage of the symmetry of the HTS module. As it can be seen in the Fig. 6.2 a), the lateral structure of the HTS module varies periodically, with a period that depends upon the number of HTS superconducting panels. If one considers an ideal case in which the properties of the material constituting the HTS module do not depend upon the angular coordinate, the information on the angular coordinate has no relevance anymore: a longitudinal slice of a period of the HTS module would be undistinguishable from any other. In a real HTS module the hypothesis of angular independence does not hold completely; indeed, the manufacturing process always involves some inhomogeneity within the required quality standards, despite its accuracy. Some examples for the HTS module are: the soldering of the HTS superconducting panels to the copper end caps and the stainless steel shunt; the assembling of the stacks from the BSCCO tapes; the sintering of the BSCCO tapes into stacks. On the other hand, in the design phase it is not possible to foresee how inhomogeneity will occur during the manufacturing process; therefore, assuming that the quality of the manufacturing process can only vary within the ranges dictated by the quality standards, the angular dependence will be neglected in this analysis.

Computational domain for the 2-D axis-symmetric model

Along their extension, the panels made of HTS superconductors typically cover a large fraction of the lateral surface. For instance, in both the W7-X HTS current leads and the 70 kA ITER Demonstrator, about 90% of the later surface spanning over the length of the HTS panels is occupied by the HTS panels themselves. In this respect, the geometry of the HTS module can be simplified as follow: instead of panels regularly displaced, the lateral surface is assumed to be entirely coated by the BSCCO superconductor. The thickness of this coating is the same of the panels' thickness. An example of HTS module under this condition is shown in Fig. 6.2 b). The computational analysis can be applied to a longitudinal section of the HTS module, which defines a 2-D axis-symmetric computational domain, as in Fig. 6.2 d).

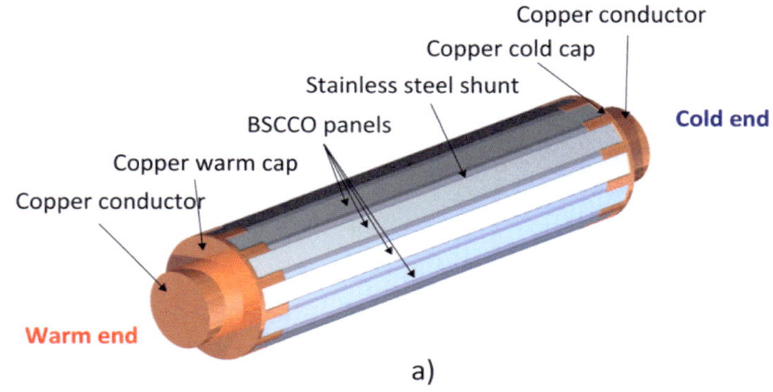

Fig. 6.2: a) Structure of a HTS module;

b) Corresponding structure of the HTS module in the 2-D axis-symmetric model approximation;

c) Corresponding structure of the HTS module in the 3-D reduced model approximation: all slices are assumed to be equal;

d) Computational domain of the 2-D axis-symmetric model;

e) Computational domain of the 3-D reduced model.

Computational domain for the 3-D reduced model

The actual structure of a HTS module can be described in more detail with a 3-D reduced model. According to the introductive considerations of this section, the HTS module can be treated as it was an electrical and thermal circuit in parallel. Each element of this circuit consists of a slice of the HTS module, as shown in Fig. 6.2 c). Each slice is equal to the others. Since a single slice is also symmetric with respect to a longitudinal symmetry plane, the computational domain for the 3-D reduced model can be reduced to half of a slice (Fig. 6.2 e)).

6.1.2 Mathematical model for the analysis

To study the design issues summarized in Tab. 6.1, it is necessary to determine the temperature and the current distribution inside the HTS module. Depending on the type of operation, i.e. normal or fault condition, the thermal and electrical analysis has to be applied in steady-state or in time-dependent form.

For the time-dependent, thermal electrical analysis, the set of equations to be solved is listed below. Equation 6.2 is the first principle of thermodynamics in the temperature-dependent form (energy conservation); Equation 6.3 is the conservation of the electrical current; Equation 6.4 is the first Ohm's law; whereas the Eq. 6.5 is the classical formulation of the electrical field $\boldsymbol{E}$ as minus the gradient of the electrical potential.

$$\rho \cdot c_{\mathrm{p}} \cdot \frac{\partial T}{\partial t} = \nabla \cdot (k \cdot \nabla T) + \dot{q}, \tag{6.2}$$

$$\nabla \cdot \boldsymbol{J} = \dot{q}_{\mathrm{j}}, \tag{6.3}$$

$$\boldsymbol{J} = \sigma \cdot \boldsymbol{E} + \frac{\partial \boldsymbol{D}}{\partial t} + \boldsymbol{J}_e, \tag{6.4}$$

$$\boldsymbol{E} = -\nabla V. \tag{6.5}$$

The set of equations for the steady-state analysis is analogous, but in this case the time derivative is set to zero and Eq. 6.2 and Eq. 6.4 simplify to Eq. 6.6 and Eq. 6.7, respectively.

$$0 = \nabla \cdot (k \cdot \nabla T) + \dot{q}, \tag{6.6}$$

$$\boldsymbol{J} = \sigma \cdot \boldsymbol{E} + \boldsymbol{J_e}. \tag{6.7}$$

Boundary conditions

Both the time-dependent and the steady-state analysis require the imposition of conditions involving the temperature T (or its derivative) and the electrical current density $\boldsymbol{J}$ on the boundary of the computational domain. This means that a condition on T and $\boldsymbol{J}$ has to be applied on each edge of the 2-D axis-symmetric computational model and on each surface of the 3-D reduced model. Moreover, the time-dependent analysis requires an initial distribution of T and $\boldsymbol{J}$ over the entire computational domain.

The boundary conditions applied to the 2-D axis-symmetric and 3-D reduced model are presented in Fig. 6.3. In steady-state, at the warm end of the HTS module a fixed temperature $T = T_{\mathrm{HTS,W}}$ and a current density of magnitude J perpendicular to the boundary are imposed (Fig. 6.3 a)). At the cold end a fixed temperature $T = T_{\mathrm{HTS,C}}$ is imposed, and the voltage V is set to zero, or at ground voltage (Fig. 6.3 b)); in this way, the conservation of the current is forced without explicitly imposing at the cold end the same value for the current density magnitude J imposed at the warm end. In Fig. 6.3 c), the symmetry conditions of the models are shown.

In the LOFA time dependent analysis, all boundary conditions are left unchanged except the condition on the fixed temperature at the warm end of the HTS module. It is substituted by imposing a time dependent temperature $T_{\mathrm{HTS,W}} = f(t)$ (Fig.6.3a)). As initial distribution of T and J, the solution of a steady-state problem is used.

All edges belonging to the 2-D axis symmetric model as well as all surfaces belonging to the 3-D reduced model that have not been explicitly treated above and highlighted in Fig. 6.3 are set as thermally and electromagnetically insulated.

The value of the current density magnitude J to be imposed at the warm end of the HTS module has to be calculated from the electrical current I flowing into the HTS module and the cross section of the copper conductor (see Fig. 6.2 a)).

In this work, the dependence of the warm end temperature $T_{\mathrm{HTS,W}}$ upon the time has been derived from the experimental results of the 70 kA ITER HTS current lead Demonstrator and W7-X HTS current lead prototype (see section 6.2) during a LOFA.

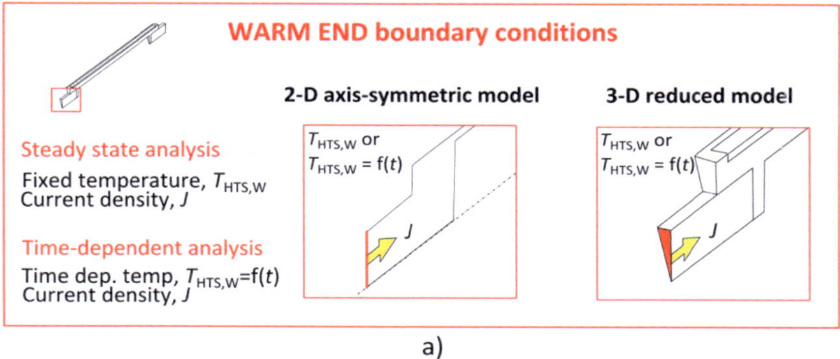

a)

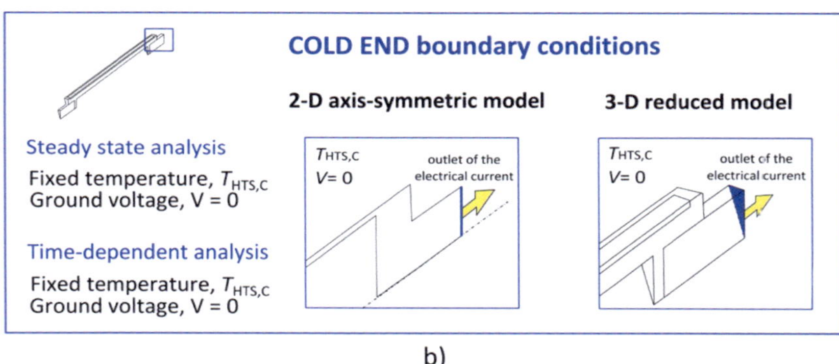

b)

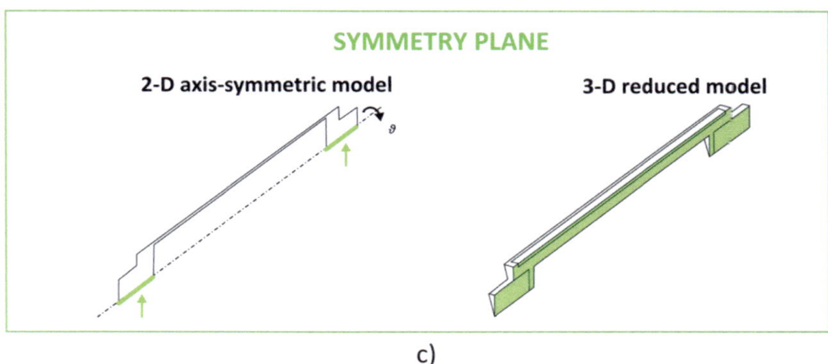

c)

Fig. 6.3: a) Boundary conditions to be imposed at the warm end of the 2-D axisymmetric and 3-D reduced model in the steady-state and time dependent-analysis;

b) Boundary conditions to be imposed at the cold end of the 2-D axisymmetric and 3-D reduced model in the steady-state and time-dependent analysis;

c) Symmetry boundary conditions for the 2-D axisymmetric and 3-D reduced model (for both steady-state and time-dependent analysis).

6.1.3 Software, numerical method and mesh

The computational domains of the 2-D axis-symmetric and 3-D reduced models have been created with the commercial software Autodesk Inventor Professional 2012®. The numerical analysis of the models has been implemented in the commercial software COMSOL Multiphysics®.

The software COMSOL Multiphysics® solves sets of partial differential equations with the Finite Elements Method.

The 2-D axis-symmetric computational domains have been discretized with a mixed triangular-rectangular mesh; the 3-D reduced computational domains have been discretized with an analogous technique, but in three-dimensions: the mesh consists of tetrahedron and rectangular prisms. A mesh made of these two discretization units (i.e. triangles/tetrahedron and rectangles/rectangular prisms) is particular suited for the HTS module geometry. Indeed, the HTS module contains regions with similar extension in the two-three dimensions, as the copper ends, but also slender regions, as the HTS panel and the stainless steel shunt. In the first case, the triangles/tetrahedron discretization can be used, whereas in the second the rectangles/rectangular prisms one. Results computed on this kind of computational grids have been tested against grid independence, as discussed in Appendix D.

6.1.4 Analysis with the 2-D axis-symmetric and 3-D reduced models

The procedure for the numerical analysis of the HTS module with the 2-D axis-symmetric and 3-D reduced model is as follow:

- in the first place, a steady-state problem is solved,
- then, the time-dependent simulation of a LOFA accident is computed.

The steady-state solution provides the temperature and current distributions within the computational domain. From these results, it is possible to calculate the cooling power P_{HTS} (Eq. 6.1) and the transfer length at the interfaces normal/ superconductor. Furthermore, the solution is also used as initial condition of the time-dependent analysis.

The time-dependent simulation provides the evolution over a prescribed time interval of the initial temperature and current distribution once the warm end temperature of the models varies according to $T_{HTS,W} = f(t)$.

6.2 Validation against experimental results

The 2-D axis-symmetric and 3-D reduced models have been applied to the analysis of the 70 kA ITER "Demonstrator" and to the W7-X HTS current leads. The aim is to compare the numerical results of the two models and to validate them against the available experimental data for both steady-state and time-dependent cases.

HTS module of the 70 kA ITER "Demonstrator"

The 70 kA ITER Demonstrator is the first prototype of HTS current lead for nuclear fusion applications [HAA04]. It has been designed, partially built and assembled at the former Forschungszentrum Karlsruhe, presently KIT, to prove the feasibility and the advantages of HTS current leads for large superconducting magnet systems in nuclear fusion reactors [HDD05]. The experimental campaign has been divided in three phases during which the Demonstrator has been tested in the electrical current range $I = 50 - 80$ kA, in nominal and fault conditions [FZK04, FZK05].

The HTS module of the Demonstrator was manufactured by the company American Superconductors. It contains 12 superconducting panels made out of 7 stacks each. The stacks contain 12 BSCCO tapes each (1008 BSSCO tapes in totals). Figure 6.4 a) shows the HTS module after the manufacture, whereas the Fig. 6.4 b) shows a particular of the 3-D reduced model computational domain and the mesh.

HTS module of the W7-X HTS current leads

The HTS module of the W7-X HTS current leads contains a lower number of BSCCO tapes with respect to the HTS module of the Demonstrator. The maximum design current is 18.2 kA and the stray field is 90 mT, therefore 360 BSCCO tapes have been used [FHK09]. As for the Demonstrator, the superconducting tapes have been organized in 12 panels made out of 5 stacks each. Each stack contains 6 tapes.

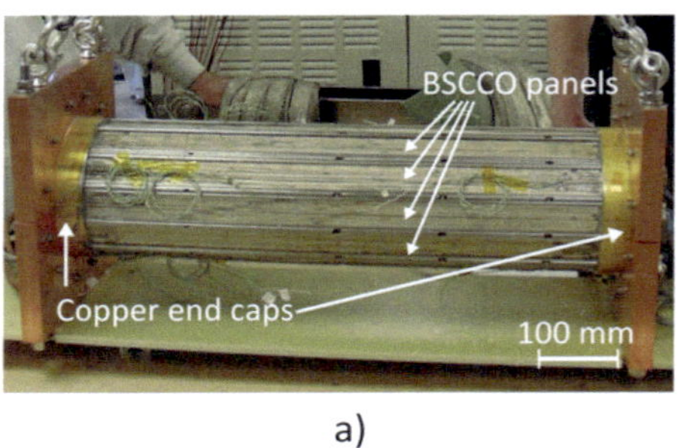

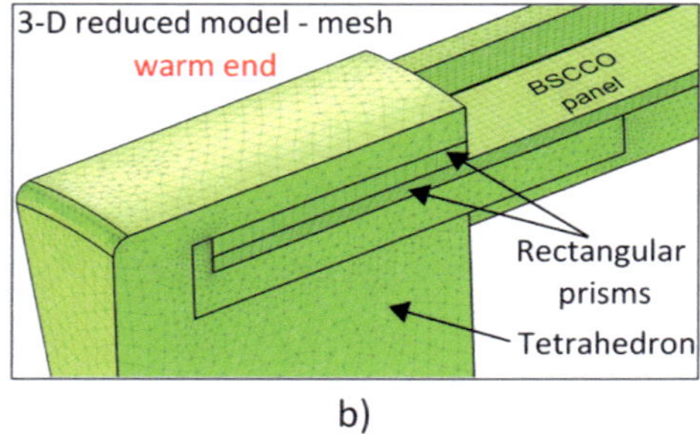

a) b)

Fig. 6.4: a) HTS module of the 70 kA Demonstrator. Source: American Superconductors;
b) Sketch illustrating the warm end of the 3-D reduced model of the 70 kA Demonstrator HTS module and the mesh composed of tetrahedron and rectangular prism.

The matrix of the tapes consists of a silver-gold alloy instead of only silver. This is of great importance, because the AgAu alloy has lower heat conductivity than the silver [SWH09]; the conducted heat flux is therefore lower even though other parameters are left unchanged (i.e. temperature gradient, geometry), as can be seen in Eq. 6.1.

Material properties

The properties of interest of all materials used in the computational analysis of the HTS module are discussed in Appendix C.

6.2.1 Steady-state analysis

The validation of the 2-D axis-symmetric and 3-D reduced model is here presented with respect to the HTS module of the W7-X HTS current lead prototype.

The temperature profile along the HTS module has been computed for electrical current I equal to I = 0, 14, 18.2 and 20 kA. At the warm and cold end of the models, experimental values of the temperature have been applied. The results are shown in Tab. 6.2. The cold end of the HTS module corresponds at the axial position 0%, whereas the warm end at the axial position 100%. The sensors arrangement of the experimental campaign [FDF11, HDF11b] allows the computed temperature profiles and experimental data to be compared at the mid-length of the HTS module (50%). As it is possible to see, both the 2-D axis-symmetric and the 3-D reduced model predict the temperature at the mid-length of the HTS module within 1 K of margin with respect to the experimental data, i.e. with a maximum relative error e equal to 3%.

Also the computed voltage drop along the HTS module is in good agreement with respect to experiments. Results for electrical current I equal to I = 14, 18.2 and 20 kA are gathered in Fig. 6.5. It is worth mentioning that the simplifications introduced with the 2-D axis-symmetric model lead to a slightly larger computed voltage drop with respect to the 3-D reduced model.

The comparison among the transfer lengths predicted with the 2-D axis-symmetric and the 3-D reduced model has been performed by plotting the normalized electrical current in the BSCCO panel I_{BSCCO}/I against the length of the panel itself. The electrical current is rated at I = 18.2 kA. As it can be seen in Fig. 6.6, the predictions of the two models are equivalent.

According to the comparison with experimental data, both the 2-D axis-symmetric and the 3-D reduced model can reproduce with good accuracy the steady-state condition of the HTS module. Since the choice of the modelling technique is not critical for the reliability of the results, also the CPU time required to prepare and solve the models should be considered. Indeed, the preparation and solution of the 2-D axis-symmetric model is faster than the 3-D reduced model. In particular, solving a 2-D axis-symmetric model typically requires about 100 s, whereas the 3-D reduced model about 1000 s.

Tab. 6.2 Steady-state temperature values computed with the 2-D axis-symmetric and the 3-D reduced model for electrical current I = 0, 14, 18.2 and 20 kA. Both models predict the temperature at the mid-length (50%) of the HTS module within 1 K difference with respect to the experimental data.

El. current, I / kA	**Experimental $T_{HTS,50\%}$ / K**	**2-D ax. sym. $T_{HTS,50\%}$ / K**	**$e_{Exp./2\text{-}D}$ / %**	**3-D red. $T_{HTS,50\%}$ / K**	**$e_{Exp./3\text{-}D}$ / %**
0	39.1	38.8	0.8	39.6	1.3
14	38	38.4	1.1	39.2	3
18.2	38.7	38.0	1.8	38.8	0.3
20	36.7	36.2	1.4	36.9	0.5

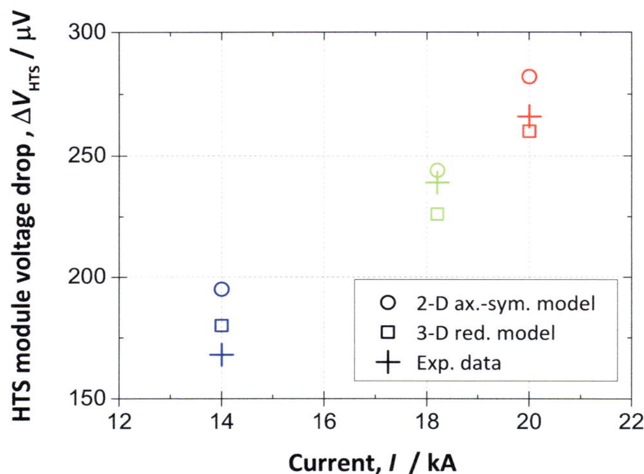

Fig. 6.5: Steady-state voltage drop over the HTS module computed with the 2-D axis-symmetric and the 3-D reduced model for electrical current I = 14, 18.2 and 20 kA. Both models predict a voltage drop in good agreement with respect to the experimental data.

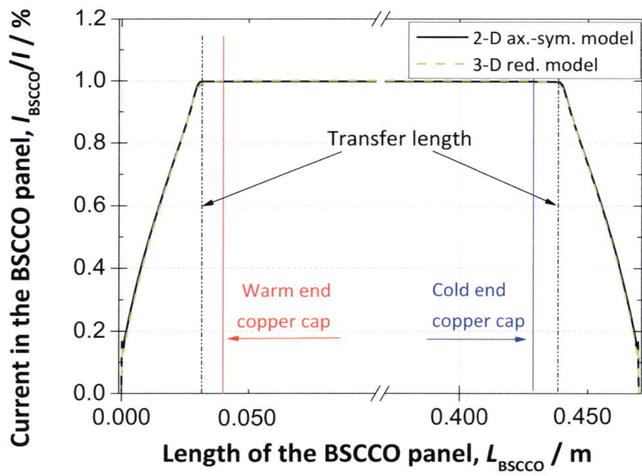

Fig. 6.6: Steady-state electrical current distribution along the BSCCO panel computed with the 2-D axis-symmetric and the 3-D reduced model. The electrical current in the BSCCO panel has been normalized with respect to the current I, which is rated at I = 18.2 kA. As it is possible to see, the contact length along which the electrical current flows into the superconductor is equivalent for the two models.

6.2.2 Time-dependent analysis

The time-dependent analysis consists in simulating a LOFA accident from its occurrence until the quench of the HTS conductor. The computational outcomes are then compared with experimental results. The 2-D axis-symmetric and the 3-D reduced models have been applied to the study of a LOFA occurring at the 70 kA ITER Demonstrator when the transported current is I = 68 kA and at a W7-X HTS current lead when the transported current is I = 18.2 kA.

As mentioned in § 6.1.2, this kind of analysis requires an initial condition on both the temperature distribution and voltage throughout the computational domain; at the warm end of the HTS module, the imposition of a time-dependent temperature boundary condition $T_{\mathrm{HTS,W}} = f(t)$ is also required (see Fig. 6.3).

As initial condition, the steady-state solution is used.

Regarding the boundary condition $T_{\mathrm{HTS,W}} = f(t)$, it has been imposed in the form of linear temperature increment, as shown in Eq. 6.8:

$$T_{\mathrm{HTS,W}} = a + b \cdot t, \tag{6.8}$$

where a is the steady state temperature at the warm end of the HTS module and b is the velocity at which this temperature increases during the transient. Values for a and b used in the time-dependent analysis are gathered in Tab. 6.3. In more detail, the parameter b has been derived from the available experimental data of the two case studies, i.e. from [FZK05, p. 27] for the 70 kA ITER Demonstrator and from [HDF11b, Hel13a] for the W7-X HTS current lead. It is worth mentioning that the

Tab. 6.3: Parameters for the time-dependent evolution of the temperature at the warm end of the HTS module during a LOFA (Eq. 6.8).

	a / K	**b / K/s**
W7-X	65.5	0.03165
70 kA Demonstrator	60	0.108

location of the boundary at the warm end of the computational models does not always reproduce completely the displacement of the sensors in the experimental set-up. Nevertheless, the difference is quite small, as can be seen in Fig 6.7.

The computed time-dependent trends of temperature and voltage drop over the HTS module for the W7-X case and of the temperature for the 70 kA Demonstrator case are shown in Fig. 6.7, Fig. 6.8 and Fig. 6.9, respectively.

As it can be seen, if the material properties are set as reported in the Appendix C, both the 2-D axis-symmetric and the 3-D reduced model predict a faster temperature increase than the experimental one. Under these conditions, the computed quench occurs between 10% and 20% in advance and with a sharper transition with respect to the experiments.

Considering the good agreement between the computed and the experimental results in the steady-state analysis, causes for the discrepancy in the time-dependent results can be related to:

- heat capacities, or masses, that are not considered in the computational models nor influence the steady-state solution, but that exist and adsorb part of the thermal energy during a transient occurring in an actual HTS module,
- heterogeneities along the angular-coordinate in actual HTS modules,
- uncertainties in the material properties implemented in the models, which are relevant for the time dependent analysis, as the specific heat capacity, the density and the resistivity of the BSCCO stacks/panels.

The first two points represents intrinsic limitations of both the 2-D axis-symmetric and the 3-D reduced model. On the other hand, the material properties are inputs given to the models themselves: their accuracy abstracts from the models' simplifying assumptions. In more detail, material properties for copper and stainless steel have been intensively studied and are presently well known. This is not the case for the specific heat capacity and density of the BSCCO stacks. Though, the most relevant quantity for triggering the quench is the BSCCO electrical conductivity: a characterization as accurate as possible of the temperature dependence of the electrical conductivity may improve the accuracy of the numerical modelling.

However, both 2-D axis-symmetric and 3-D reduced model predict the occurrence of the quench in advance with respect to the experiments: their prediction is therefore pessimistic, hence conservative.

Although the accuracy of the computed results can be improved, both models can be used for a conservative analysis of the transient occurring as a consequence of a LOFA accident.

As for the steady-state analysis, differences in the results computed with the two models are limited. The 2-D axis-symmetric model might be therefore preferred, since its solution requires a shorter CPU time than the solution of the corresponding 3-D reduced model.

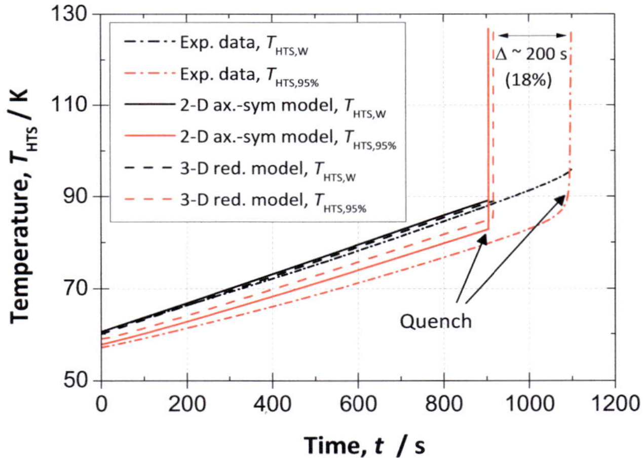

Fig. 6.7: W7-X case: temperature increase during a LOFA accident with current I rated at I = 18.2 kA. The temperature $T_{HTS,W}$=f(t) has been imposed at the warm end of the 2-D axis-symmetric and the 3-D reduced model as indicated in Eq. 6.7. The computed values at the 95% of the HTS module length from the cold end are compared with the experimental data at the same position. The computed temperature increases faster and the quench is detected about 18% in advance with respect to the experimental data. The computed results also show a sharper trend than the experiment.

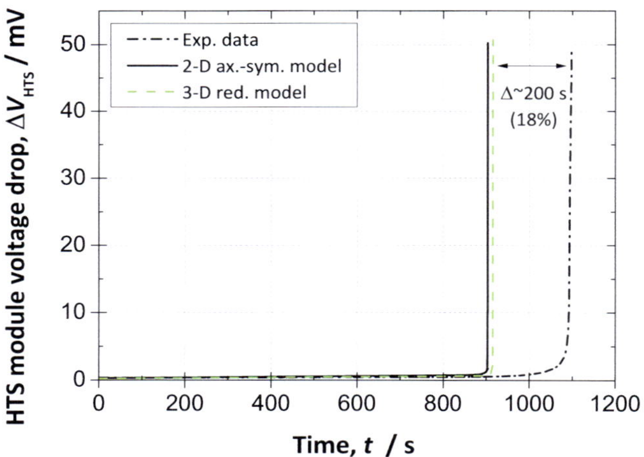

Fig. 6.8: W7-X case: voltage drop over the HTS module during a LOFA accident with current I rated at I = 18.2 kA. The computed voltage drop over the HTS module shows the quench to occur about 18% in advance with respect to the experimental data.

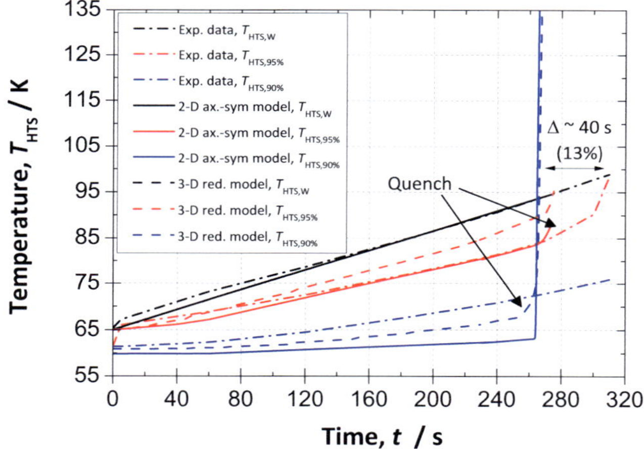

Fig. 6.9: 70 kA Demonstrator case: temperature increase during a LOFA accident with current I rated at I = 68 kA. The temperature $T_{HTS,W}$=f(t) has been imposed at the warm end of the 2-D axis-symmetric and the 3-D reduced model as indicated in Eq. 6.7. The computed values at 90% and 95% of the HTS module length from the cold end are compared with the experimental data at the same position. The computed temperature increases faster and the quench is detected about 13% in advance with respect to the experimental data. The computed results also show a sharper trend than the experiment.

6.3 Summary

In this Chapter, numerical models have been proposed for the analysis of the HTS module in steady-state (normal operation) and time-dependent (LOFA accident) conditions. The computed outcomes have been compared against experimental results with the purpose of validation.

Taking advantage of the typical structure of a HTS module, two types of model have been considered:

- a 2-D axis-symmetric model, and
- a 3-D reduced model.

Both the 2-D axis-symmetric and the 3-D model show a good capability in predicting the steady-state operation of the HTS module; on the other hand, both models predict a more pessimistic transient during a LOFA accident with respect to the experiments (10-20 % in advance). Although this can be due to the simplifications introduced with the numerical models themselves, it might be more likely related to the accuracy of the material properties implemented therein and in particular to the temperature dependence of the electrical conductivity of the superconductor. More accurate material properties may therefore improve the time-dependent predictions of both the 2-D axis-symmetric and the 3-D reduced model.

However, since the models predict a more severe transient than the experimental data, they can be used for a conservative analysis of a LOFA accident.

The computed results also show that there is no clear advantage in choosing between the 2-D axis-symmetric and the 3-D reduced model. The former one can be therefore preferred since its solution (for both a steady-state and a transient analysis) requires a much shorter CPU time than the solution of a corresponding 3-D reduced model.

7 Predictive analysis of ITER HTS current leads

In this Chapter, a predictive analysis and independent verification of the ITER HTS current leads' design is presented. To the author's knowledge, a full length analysis of ITER HTS current leads is not yet available in the literature. The correlations presented in Chapter V are here applied to the study of the meander flow heat exchangers mounted in ITER HTS current leads. The results obtained are compared with the ITER relevant requirements for the HTS current leads. The content of this Chapter has been published by this author in [RHS13c].

7.1 ITER HTS current leads

ITER HTS current leads will feed the ITER magnet system. The ITER magnet system [MDL12] consists of:

- eighteen Toroidal Field, TF, coils,
- six Poloidal Field, PF, coils,
- six Central Solenoid, CS, modules,
- nine Correction coils, CC, pairs.

A bird-eye view of the superconducting magnets arrangement is shown in Fig. 7.1.

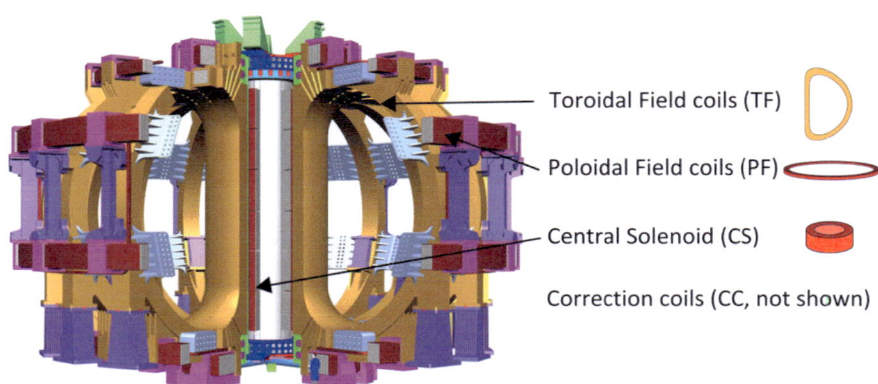

Fig. 7.1: ITER magnet system.

Three kinds of HTS current leads will be used to feed the ITER magnet system:

- for the TF coils, 18 HTS current leads carrying a maximum current of 68 kA,
- for the PF coils /CS, 24 HTS current leads carrying a maximum current of 55 kA,
- for the CC coils, 18 HTS current leads carrying a maximum current of 10 kA.

Although each of these HTS current leads is rated at a different maximum current, their design is similar. This design approach was proven to offer high reliability and a consistent reduction of cooling power, with respect to a conventional current lead, for both nuclear fusion and high energy particle physics research applications [HAA04, BMM03, HFK11]. A longitudinal section of the HTS current leads for the TF coils is shown in Fig. 7.2. The main features of this HTS current lead type are conceptually representative of the PF/CS and CC HTS current leads as well.

The connection to the room temperature power supply is realized at the copper room temperature terminal, which is equipped with its own heat exchanger. The room temperature terminal is connected, in correspondence of the flange, to the copper meander flow heat exchanger. Both these components need to be actively cooled by gaseous helium. On the opposite (viz. cold) side, the meander-flow heat exchanger is connected to the HTS module, which is made of a stainless steel shunt provided with copper ends. On the outer surface of the module BSCCO-2223, Bi-2223, panels are housed in longitudinal slots.

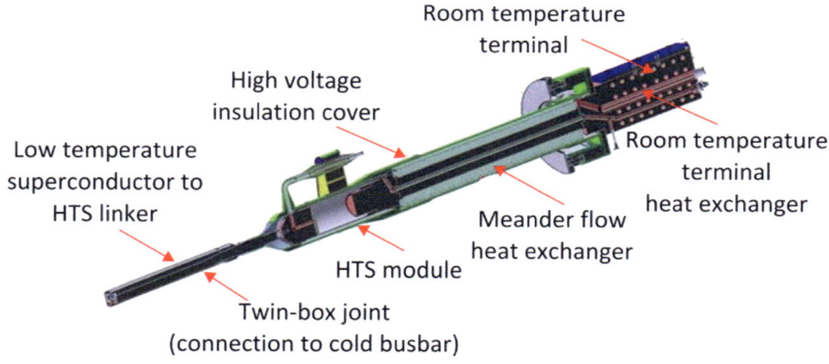

Fig. 7.2: Longitudinal section of a 68 kA HTS current lead for the TF coils of the ITER magnet system [Bau12].

For the TF leads, the module will be provided with 90 slots, each for a stack of 12 Bi-2223 tapes. The HTS modules for the PF/CS leads will also have 90 slots, but each of the stacks contains 10 Bi-2223 tapes. In the CC leads there are 36 slots with stacks containing 4 tapes each. The HTS module is expected to operate in the temperature range 5-65 K and it is cooled by conduction from the cold end. The cold copper end of the HTS module is finally connected via the twin-box joint to the cold busbar.

Gaseous helium will enter the HTS current leads at the transition between the HTS module and the meander flow heat exchanger and flow towards the warm end. The foreseen helium inlet conditions are $T_{He,in}$ = 50 K and $p_{He,in}$ = 0.3 MPa. A significant peculiarity of the ITER HTS current lead design is the extension of the helium cooling to the room temperature terminals [BBB12, Tay12]. Indeed, heat exchangers obtained by wire cutting longitudinal fins in a copper rod will be housed inside the rectangular copper terminal blocks.

Figure 7.3 shows a cross section of heat exchanger inside the room temperature terminal for the TF HTS current lead case.

A selection of the design requirements of the ITER HTS current leads relevant for the modelling presented in this work has been outlined in Tab. 7.1.

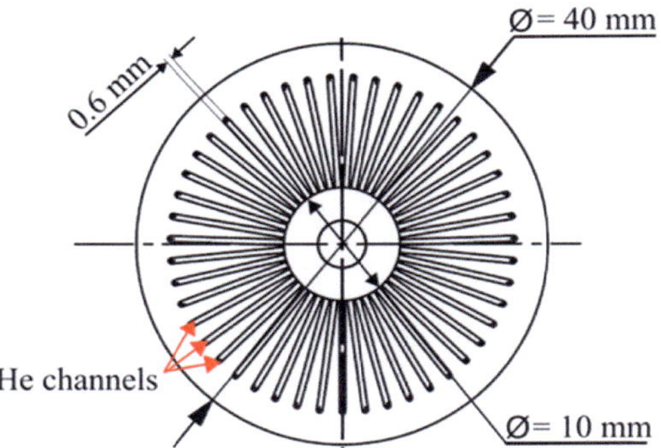

Fig. 7.3: Cross section of the heat exchanger housed in the room temperature terminal of the TF HTS current leads. The helium flows inside rectangular channels (perpendicular to the plane of the section) obtained by wire cutting a copper rod in the longitudinal direction. The central part of the heat exchanger is filled.

Tab. 7.1: Design requirements of the HTS current leads for ITER. Source [BBB12, Bau12, Tay12].

	TF, 68 kA	PF/CS, 55 kA	CC, 10 kA
Operating temp. of the HTS module / K	5-65	5-65	5-65
HTS contact resistance at 65 K, $R_{HTS\text{-}HX}$ / nΩ	≤10	≤10	≤10
HTS contact resistance at 5 K / nΩ	≤1	≤1	≤1
Operating temp. meander flow heat exchanger/ room temperature terminal / K	65 - 300	65 - 300	65 - 300
Helium temp. at meander flow heat exchanger inlet, $T_{He,in}$ / K	50±1	50±1	50±1
Maximum helium mass flow rate, $\dot{m}$ / g/s	4.8	3.85	0.7
Maximum pressure drop over the HTS current lead / MPa	0.2	0.2	0.2
Maximum heat load at the cold end. $\dot{Q}_{cold}$ / W	15	12	3

7.2 Scope and methodology of the analysis

The ITER organization, IO, is responsible for the design of the ITER HTS current leads; it is supported by the Institute for Plasma Physics of the Chinese Academy of Sciences, ASIPP, as well as by the European Centre for Nuclear Research, CERN, which played a pivotal role in the design of the resistive heat exchanger of the ITER HTS current leads [BBB12]. A series of analyses based on 3-D, 2-D and 1-D models were undertaken [Tay12] in order to find optimized designs of the three ITER HTS current lead types. These analyses led to the present designs (see [BBB12]), which should minimize the cooling power needed to properly operate the HTS current leads. However, the overall behaviour of a HTS current lead does not depend solely upon the performance of its single components, but also on their mutual interactions. A predictive analysis of the overall performance of the ITER current leads would therefore be worthwhile; nevertheless, to the best of the author's knowledge, it is not yet available in the literature.

This analysis aims at filling the above-mentioned lack of knowledge about the full-length HTS current leads' behaviour. Full length 1-D models of the three types of ITER HTS current leads have been implemented in the code CURLEAD [Hel89] and a steady-state, thermal-hydraulic analysis of their normal operative conditions has been performed.

The correlations derived in Chapter V have been used to model the thermal-fluid dynamics of the helium inside the meander flow heat exchanger.

The computed results have been compared in the last sections with the ITER requirements relevant for the HTS current leads.

7.2.1 CURLEAD code

The code CURLEAD [Hel89] solves the 1-D conjugate heat transfer problem for the HTS current lead and the coolant. The equations required for the steady state analysis are described hereunder. The heat equation for a current lead is:

$$A \cdot \frac{d}{dx}\left(\lambda(T) \cdot \frac{dT}{dx}\right) - h \cdot P \cdot (T - T_{\mathrm{He}}) + A \cdot \frac{J^2}{\sigma(T)} = 0, \tag{7.1}$$

where λ(T) is the heat conductivity, A the cross section of the lead, h the heat transfer coefficient, P the cooling perimeter, T_{He} the temperature of the coolant, σ(T) the electrical conductivity and J the current density.

To model the coolant, helium in this case, the transport equation for the energy and an equation for the pressure drop are needed:

$$\dot{m}_{\mathrm{He}} \cdot c_{\mathrm{p,He}} \cdot \frac{dT_{\mathrm{He}}}{dx} - h \cdot P \cdot (T - T_{\mathrm{He}}) = 0, \tag{7.2}$$

$$\Delta p_{\mathrm{He}} = \zeta \cdot \frac{\dot{m}_{\mathrm{He}}^2}{2 \cdot \rho_{\mathrm{He}} \cdot A_{\mathrm{He}}}, \tag{7.3}$$

where $\dot{m}$ is the helium mass flow rate, $c_{\mathrm{p,He}}$ the helium specific heat capacity, ζ is the pressure drop coefficient, ρ_{He} is the helium density and A_{He} is the cross section characterizing the coolant's flow.

Both the heat transfer and the pressure drop coefficient depend on the flow condition, i.e. on the thermodynamic state of the coolant and on the geometry of the channel. In the code, they are provided in the form of correlations depending on the

Reynolds number, *Re*. The solution of Eq. 7.1 requires a boundary condition on both boundaries of the 1-D domain; CURLEAD can handle either a couple of Dirichlet type boundary conditions, i.e. a fixed temperature, or a combination of Dirichlet on one boundary and Neumann, i.e. a condition on the temperature gradient, on the other. Equations 7.2 and 7.3 require an inlet condition for the helium temperature and for the pressure, respectively.

Equations 7.1-7.3 are discretized with the finite difference method and solved with the segregated approach.

7.2.2 1-D models of ITER HTS current leads

For the purpose of this work, the most interesting section of a HTS current lead spans from the room temperature terminal down to the cold copper end of the HTS module (see Fig. 7.2). Indeed, in steady state operation, the largest temperature gradient occurs over this length, whereas departing from the copper cold end of the HTS module the temperature variation is lower than 1 K. The 1-D models of the ITER HTS current leads cover therefore the length from the room temperature terminal down to the cold copper end of the HTS module.

The models have been created by reducing the geometry of each part of a HTS current lead into a one-dimensional, i.e. longitudinal, element. Each of these elements is characterized by its length, cross sections (A in Eq. 7.1), amount of carried current, composition in terms of material and cooling condition. Regarding the materials, for the Bi-2223 stacks the properties of those used for W7-X have been implemented [KSW09, HFK08], whereas copper with RRR = 50 has been adopted. The contact resistance at the cold end of the HTS module has been assumed to be 1 nΩ. The value of the contact resistance at the 65 K end of the HTS module, $R_{HTS\text{-}HX}$, has been varied parametrically in the range 1-10 nΩ, for it influences the heat generation in this region quite relevantly, affecting therefore the demand for cooling power.

The helium cooling circuit inside the model of the HTS current leads consists of a series of two heat exchangers: the heat exchanger obtained by wire cutting longitudinal fins in a copper rod that is housed inside the rectangular copper terminal blocks, and the meander flow heat exchanger. A schematic view of the models implemented in

CURLEAD is shown in Fig. 7.4 (the solid line of the helium cooling circuit refers to the meander flow heat exchanger, whereas the dashed-dotted line to the room temperature terminal heat exchanger).

The 1-D modelling of the meander flow and the room temperature terminal heat exchangers requires the knowledge of some geometrical parameters (the hydraulic diameter d_h, the helium cross section A_{He} and the cooling perimeter P) and correlations for the friction factors (or pressure drop coefficient) and heat transfer coefficients characterizing the helium thermal-hydraulics along the heat exchangers. Considering the meander flow heat exchanger, definitions for $d_{h,MF}$ and $A_{He,MF}$ in the meander geometry have been provided in Chapter V along with the correlations for both the pressure drop coefficient, ζ, and the Nusselt number, Nu, depending on the Reynolds numbers, Re. The characteristic geometrical parameters of the meander flow heat exchanger for the TF, PF/CS and CC HTS current leads are presented in Tab. 7.2, whereas the coefficients of the corresponding correlations implemented in CURLEAD are shown in Tab. 7.3: for $Re < 1000$ and $Re > 2000$ the correlations are in the form $\zeta = A \cdot Re^B$ and $Nu = C \cdot Re^D$, whereas for Re 1000-2000 in the form $\zeta = A + B \cdot Re$ and $Nu = C + D \cdot Re$.

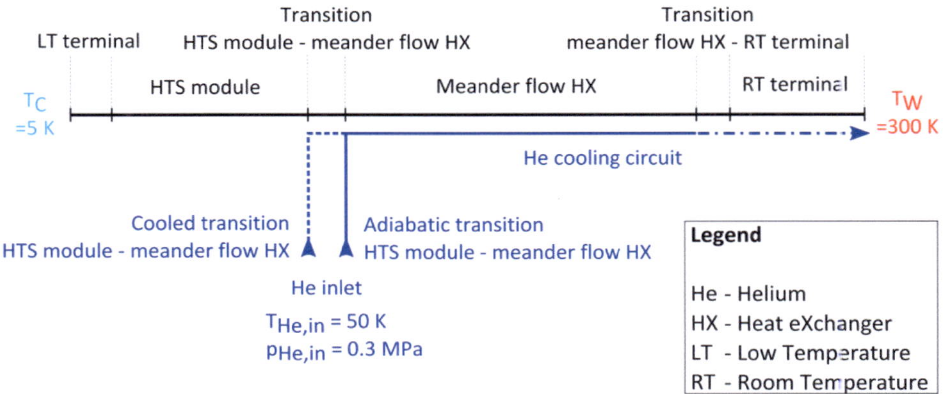

Fig. 7.4: Schematic view of the helium circuit as it is modelled with CURLEAD. The helium enters the resistive part of the HTS current lead either at the cold side of the transition HTS module-meander flow heat exchanger (dashed line, for the models with cooled transition), or at the cold side of the meander flow heat exchanger (solid line, for the models with adiabatic transition). Inside the current lead, the helium circuit consists of a series of two heat exchangers: the meander flow heat exchanger (solid line) and the heat exchanger housed in the room temperature terminal (dashed-dotted line).

Tab. 7.2: Characteristic geometrical quantities of the meander flow heat exchangers in the TF, PF/CS and CC HTS current leads ($d_{h,MF}$ calculated with Eq. 5.7; $A_{He,MF}$ calculated with Eq. 5.3).

	$d_{h,MF}$ / m	$A_{He,MF}$ / m^2	P_{MF} / m^2/m
TF	5.8e-03	2.87e-04	6.85
PF/CS	5.8e-03	2.74e-04	6.00
CC	5.7e-03	1.90e-04	2.67

Tab. 7.3: Coefficients for the correlations of the meander flow heat exchangers. For $Re < 1000$ Eqs. 5.16 and 5.25 have been used; whereas for $Re > 2000$, Eqs. 5.14 and 5.24. The coefficients for the cases $1000 < Re < 2000$ have been derived by linearly interpolating values for the pressure drop coefficient ζ and the Nusselt number Nu at $Re = 1000$ and $Re = 2000$.

		A	**B**	**C**	**D**
	$Re < 1000$	197.26	-0.52	5.4	0.10
TF	$1000 < Re < 2000$	6.10	-0.000668	11.32	-0.000553
	$Re > 2000$	11.93	-0.12	0.04	0.73
	$Re < 1000$	194.58	-0.52	5.54	0.10
PF / CS	$1000 < Re < 2000$	5.81	-0.000449	11.73	-0.000685
	$Re > 2000$	12.22	-0.12	0.04	0.73
	$Re < 1000$	160.18	-0.52	6.23	0.10
CC	$1000 < Re < 2000$	4.01	0.000399	12.95	-0.000709
	$Re > 2000$	11.98	-0.12	0.045	0.73

The heat exchanger housed in the room temperature terminal has been modelled as a hydraulic circuit made of 47 (TF, PF/CS cases) or 30 (CC case) parallel ducts with rectangular cross section. Under the assumption of flow homogeneously distributed among the different channels, the resultant helium flow cross section ($A_{He,RT}$) for these heat exchangers is simply given by the sum of the cross sections of each rectangular channel. A formulation for the hydraulic diameter ($d_{h,RT}$) and correlations for the thermal-hydraulics in rectangular-shaped ducts can be found in the literature (see for instance [KSA87]). The heat transfer in these heat exchangers has been modelled in analogy with rectangular ducts having one short edge kept adiabatic (this assumption is justified by the geometrical arrangement of these heat exchangers [BBB12]). Beside the meander flow and the room temperature terminal heat exchangers, two other components shall be considered in the helium cooling circuit of the HTS current leads: the transition between the HTS module and the meander flow heat exchanger, where the gaseous helium enters the current leads, and the connection between the meander-flow heat exchanger and the room temperature terminal. Starting with the latter one, it has been verified that neglecting the helium cooling in this region does not lead to any major change in the computed results (both the heat transfer surface and the length of the channel in which the helium flows are small). On the other hand, cooling the transition between the HTS module and the meander-flow heat exchanger plays a much more important role, in particular because of the ohmic heating at low temperature (~65-70 K) and the presence of the contact resistance discussed above. Due to the complexity of the helium flow in this region and to inlet effects, an exhaustive description of the coolant thermal-fluid dynamics is not trivial and beyond the scope of this analysis. For this reason, the following, simplified cases have been considered: in the first place, the transition between HTS module and meander flow heat exchanger has been treated as adiabatic (solid helium inlet line in Fig. 7.4); then as it was cooled by the inflowing helium (dashed helium inlet line helium inlet in Fig. 7.4). To define, at least qualitatively, the effects of the cooling, cooling models for parallel (par. flow) and cross flow (cro. flow) have been used. In the first case, the helium has been assumed to flow into the transition parallel to the axis of the current lead, whereas in the second perpendicular to it. The heat transfer coefficient for the parallel flow case has been derived from the well-known Dittus-Boelter correlation [KSA87]. For the cross flow case, the correlation for the meander-flow geometry has been used. No pressure drop has been considered for both cooled transition cases (i.e. par. flow and cro. flow) since its contribution to the total pressure drop occurring over the helium circuit is negligible.

7.2.3 Procedure of the 1-D analysis

The present work aims at assessing the steady state performance of the ITER HTS current leads in normal operative conditions. The design of the current leads is fixed, whereas the contact resistance $R_{\text{HTS-HX}}$ as well as the cooling condition of the transition between HTS module and meander flow heat exchanger are treated as variable parameters.

The general procedure consists of solving a 1-D model with CURLEAD for each of the ITER HTS current lead type and varying either the contact resistance or the cooling condition of the transition HTS module-heat exchanger or both. For each case, the input helium mass flow rate $\dot{m}$ is tuned in order to keep the HTS module warm end at 65 K (as specified in Tab. 7.1, first row). At the boundaries of the 1-D models, Dirichlet boundary conditions have been imposed: $T_C = 5$ K and $T_W = 300$ K. As inlet conditions for the helium, $T_{He,in} = 50$ K and $p_{He,in} = 0.3$ MPa have been used. As shown in [BBB12, Tay12], additional, external heat fluxes can be provided to the room temperature terminal of the HTS current leads by means of heaters (TF, PF/CS and CC HTS current leads) or water heat exchangers (TF, PF/CS HTS current leads). Nevertheless, none of these contributions have been taken into account since they represent optional mitigation measures used to shape the temperature profile along the HTS current leads.

7.3 Results of the analysis

The computed results are provided with error bars, which are evaluated from the error bars of the correlations presented in Chapter V.

7.3.1 Influence of the contact resistance

It is well known (see, for instance, [Wils83, p. 256]) that the main resistive components of a HTS current lead (i.e. the room temperature terminal and the meander-flow heat exchanger) can be designed in order to balance, for a specific value of the electrical current, the Joule and the conducted heat, thus minimizing the heat leak of the HTS current lead. Under the assumptions introduced in § 7.2.3, the minimization of the heat leak has to be intended as $\dot{Q}_W = 0$, where $\dot{Q}_W$ is the heat load on the current lead at its room temperature end. The presence of a contact resistance at the transition between the meander flow heat exchanger and the HTS module significantly contributes to the resistive losses at the cold end of the heat exchanger.

Since the design of the ITER HTS current leads is fixed in this work, the variation of the contact resistance $R_{\text{HTS-HX}}$ directly influences the heat load $\dot{Q}_{\text{W}}$. Indeed, the Joule and the conducted heat must be balanced for a specific electrical current in order to have the heat load $\dot{Q}_{\text{W}} = 0$, but a variation of $R_{\text{HTS-HX}}$ modifies the Joule heat contribution; therefore a variation of $R_{\text{HTS-HX}}$ leads to a misbalance among the Joule and the conducted heat, or in a change in the heat load $\dot{Q}_{\text{W}}$.

Figure 7.5 shows the dependence of $\dot{Q}_{\text{W}}$ on $R_{\text{HTS-HX}}$ for the cases treated in this work. In all cases, the heat load $\dot{Q}_{\text{W}}$ linearly increases with $R_{\text{HTS-HX}}$ and the trend becomes steeper as the current transported by the lead increases. The cooling of the transition between the HTS module and the meander flow heat exchanger, for both parallel and cross flow involves smaller heat loads with respect to the adiabatic case (no cooling). It is not always possible to have the heat load $\dot{Q}_{\text{W}} = 0$ within the range of $R_{\text{HTS-HX}}$ covered in this work. For the TF HTS current lead case, the assumption of adiabatic transition between the HTS module and the meander flow heat exchanger leads to $\dot{Q}_{\text{W}} > 0$. This means that an incoming external heat flux is entering the HTS current lead from its room temperature side. On the other hand, the results of the 1-D models with cooled transition show that $\dot{Q}_{\text{W}}$ cancels out for $R_{\text{HTS-HX}} \sim 4.5$ nΩ (par. flow) and $R_{\text{HTS-HX}} \sim 5.1$ nΩ (cro. flow). The heat load $\dot{Q}_{\text{W}}$ becomes negative for smaller values of $R_{\text{HTS-HX}}$, which means that overheating is occurring at some location between the meander flow heat exchanger and the room temperature terminal. For the PF/CS HTS current leads case, $\dot{Q}_{\text{W}}$ is always positive if the transported current is 52 kA and the transition between the HTS module and the meander flow heat exchanger is adiabatic, whereas it is always negative if the lead is operated at 55 kA and the transition is cooled. In the other cases, it is possible to have $\dot{Q}_{\text{W}} = 0$ if the current is 55 kA and the transition is not cooled (for $R_{\text{HTS-HX}} \sim 6.0$ nΩ), or if the current is 52 kA and the transition is cooled (for $R_{\text{HTS-HX}} \sim 4.8$ nΩ in the case of parallel flow and for $R_{\text{HTS-HX}} \sim 5.4$ nΩ in the case of cross flow). For the CC HTS current lead case, $\dot{Q}_{\text{W}}$ is always positive.

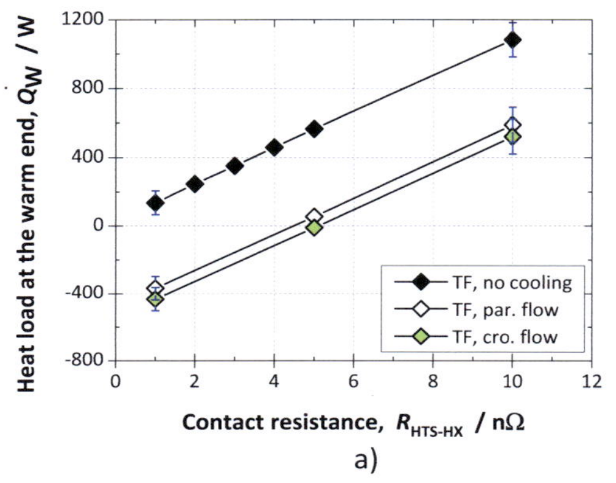

a)

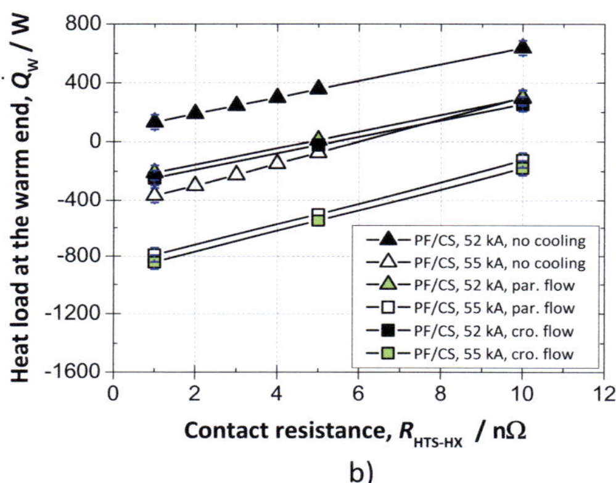

b)

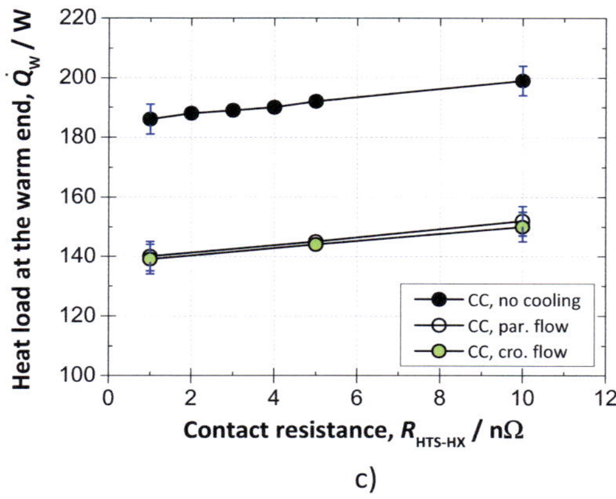

c)

Fig. 7.5: a) $\dot{Q}_W$ dependence on R_{HTS-HX} for the TF HTS current lead;
b) $\dot{Q}_W$ dependence on R_{HTS-HX} for the PF/CS HTS current lead;
c) $\dot{Q}_W$ dependence on R_{HTS-HX} for the CC HTS current lead.

Also the helium mass flow rate, $\dot{m}$, shows a linear dependence upon $R_{HTS\text{-}HX}$, although this is weaker than the $\dot{Q}_W$ one. The computed values for $R_{HTS\text{-}HX} = 1$ nΩ and 10 nΩ have been gathered in Tab. 7.4, whereas the Tab. 7.5 contains the mass flow rate and the corresponding $R_{HTS\text{-}HX}$ values for the models where it has been possible to obtain $\dot{Q}_W = 0$. According to the results presented in Fig. 7.4, some external heat source (in case $\dot{Q}_W > 0$) or heat sink (in case $\dot{Q}_W < 0$) is needed to stabilize the room temperature terminal temperature profile when $\dot{Q}_W \neq 0$. For the TF and PF/CS HTS current leads, both the $\dot{Q}_W > 0$ case and the $\dot{Q}_W < 0$ case can occur. Either the heaters or the water heat exchanger ([BBB12, Tay12]) mentioned in section 7.2.3 may therefore be needed during the normal operations of the leads. On the contrary, the design of the CC HTS current lead does not include any water heat exchanger at the room temperature terminal, but heaters only. According to the present results, this should be of no concern for the operation of the CC current leads since $\dot{Q}_W$ is always positive. The computed helium mass flow rates generally fulfil the ITER requirements except for the TF HTS current lead with adiabatic transition between the HTS module and the meander flow heat exchanger. In this case, the present analysis predicts values of $\dot{m}$ 2% larger than 4.8 g/s at most (see Tab. 7.1 and Tab. 7.4) for $R_{HTS\text{-}HX} > 8$ nΩ. Nevertheless, besides the relatively large value of $R_{HTS\text{-}HX}$, it is noteworthy recalling that the assumption of an adiabatic transition between the HTS module and the meander flow heat exchanger represents a borderline case. Indeed, according to the present design of the TF HTS current leads, the transition will actually be cooled and the results (see Tab. 7.4) show that, under this condition, the helium mass flow rate is expected to be smaller than $\dot{m} = 4.8$ g/s.

Tab. 7.4: Computed helium mass flow rates for $R_{HTS\text{-}HX}$ = 1 nΩ and 10 nΩ.

	$\dot{m}$ / g/s	
	$R_{HTS\text{-}HX}$ = 1 nΩ	$R_{HTS\text{-}HX}$ = 10 nΩ
TF lead – no cooling	4.581±0.007	4.878±0.006
TF lead – par. flow	4.445±0.013	4.703±0.014
TF lead – cro. flow	4.444±0.012	4.694±0.016
PF/CS lead, 55 kA – no cooling	3.667±0.005	3.843±0.005
PF/CS lead, 55 kA – par. flow	3.567±0.009	3.721±0.008
PF/CS lead, 55 kA – cro. flow	3.564±0.009	3.711±0.008
PF/CS lead, 52 kA – no cooling	3.443±0.004	3.599±0.005
PF/CS lead, 52 kA – par. flow	3.348±0.006	3.484±0.007
PF/CS lead, 52 kA – cro. flow	3.346±0.006	3.480±0.007
CC lead – no cooling	0.645±0.001	0.651±0.006
CC lead – par. flow	0.627±0.002	0.631±0.002
CC lead – cro. flow	0.627±0.002	0.632±0.002

Tab. 7.5: Helium mass flow rates and corresponding $R_{HTS\text{-}HX}$ values at which $\dot{Q}_W = 0$. For the cases not shown in the table, it has not been possible to obtain $\dot{Q}_W = 0$ within the $R_{HTS\text{-}HX}$ range 1-10 nΩ.

	$\dot{m}$ / g/s	$R_{HTS\text{-}HX}$ / nΩ
TF lead – par. flow	4.545±0.013	~4.5
TF lead – cro. flow	4.558±0.014	~5.1
PF/CS lead, 55 kA –no cooling	3.765±0.005	~6.0
PF/CS lead, 52 kA – par. flow	3.405±0.006	~4.8
PF/CS lead, 52 kA – cro. flow	3.412±0.007	~5.4

7.3.2 Heat load at the cold end of the current leads

The heat load at the cold end of the HTS current leads, $\dot{Q}_C$, is a relevant parameter since it defines the cooling power which has to be provided at 5 K. The results of the 1-D modelling show that $\dot{Q}_C$ is basically constant (maximum variation < 3%) for each

ITER HTS current lead type. This means that it is mainly due to the heat conducted along the HTS module and depends, therefore, just upon the materials and on the geometry, being the end temperatures fixed (at 5 and 65 K respectively). The maximum values of $\dot{Q}_C$ for each lead type are summarized in Tab. 7.6.

The computed $\dot{Q}_C$ values are below the maximum threshold of the ITER requirements for all cases.

Tab. 7.6: Computed heat loads at the cold end of the HTS current leads in normal operation.

	$\dot{Q}_C$ / W
TF lead	13.73
PF/CS lead, 55 kA	11.07
PF/CS lead, 52 kA	10.69
CC lead	1.32

7.3.3 Convective heat transfer

According to the operative conditions covered in the present work and the criteria given for the definition of the flow regime in the meander-flow geometry in Chapter V, the meander-flow heat exchanger of the TF and PF/CS HTS current leads operates in turbulent flow regime. In the first case the Reynolds number, *Re*, varies from *Re*~15000 (at 50 K) to *Re*~4300 (approaching room temperature), whereas in the second from *Re*~13000 (55 kA case, at 50 K) to *Re*~3600 (52 kA case, approaching room temperature). On the contrary, the lower helium mass flow rate shifts the CC HTS current lead operative *Re* range towards the laminar regime. Indeed, *Re* varies in this case from *Re*~3000 (at 50 K) to *Re*~1000 (approaching room temperature). For all three current lead types, the heat exchanger housed in the room temperature terminal operates in laminar regime ($Re < 1700$). The pressure drop experienced by the helium flowing along the HTS current leads is mainly concentrated in the meander-flow heat exchangers. Although the heat exchangers housed in the room temperature terminals operate at higher temperatures (i.e. higher kinematic viscosity of the helium) with

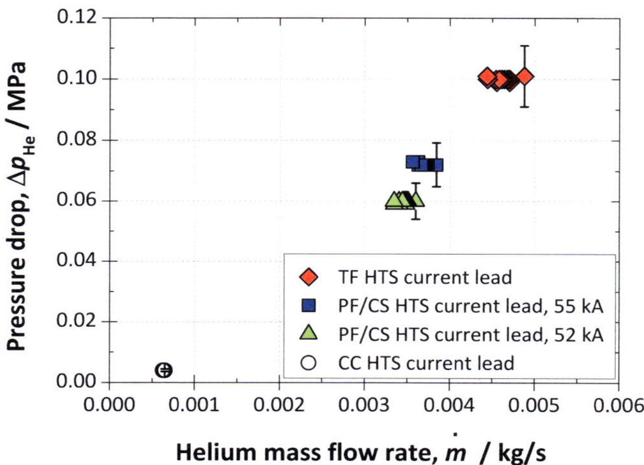

Fig. 7.6: Pressure drop occurring in the meander flow heat exchangers as a function of the helium mass flow rate.

respect to the meander-flow ones, the pressure drop occurring within them reaches at most ~2% of the total pressure drop, due to the low impedance of the rectangular channels. The computed pressure drop across the meander flow heat exchangers, Δp_{He}, has been plotted in Fig. 7.6 as a function of the helium mass flow rate for the cases covered in this work.

A comparison between the pressure drop across the HTS current lead calculated in this work and the corresponding requirements in Tab. 7.1 is unfortunately not so straightforward. The ITER specification accounts also for the pressure losses in the cryogenic circuit, including valves, flow meters, piping etc., and not only for the pressure drop across the HTS current lead itself. Presently the design of the cryogenic feeding system for the current leads is still at an early stage, therefore not even an estimation of its contribution to the overall pressure drop is available. Against this backdrop, one can in principle reverse the approach and, on the basis of the pressure drop calculated with CURLEAD, define the maximum pressure loss allowed in the cryogenic circuit. Considering that all lead types share the same pressure drop threshold (0.2 MPa) and that, according to the results, the largest pressure loss across the current lead occurs for the TF HTS current lead case, the pressure drop in the cryogenic circuit should not overcome 0.101 ∓ 0.011 MPa.

The Nusselt number calculated with the correlations presented in Chapter V is in the range 18-44 for the TF HTS current lead, in the range 16-40 for the PF/CS HTS current lead and in the range 12-17 for the CC HTS current lead. Since the helium inlet conditions are the same in all cases covered in this work and the helium specific heat capacity can be considered as constant over the range 50-300 K, the temperature difference at the end of the meander flow heat exchanger, ΔT_W, is a valid indication of the effectiveness of the convective heat transfer. Figure 7.7 shows the temperature difference between the copper and the helium at the warm end of the meander-flow heat exchanger as a function of the copper temperature, $T_{Cu,W}$, at the same location. For all cases, a reduction of $R_{HTS\text{-}HX}$ leads to higher $T_{Cu,W}$ and lower ΔT_W at the warm end. This should not be necessarily regarded as advantageous, for negative $\dot{Q}_W$ are associated with lower $R_{HTS\text{-}HX}$, as shown in Fig. 7.5. Indeed, under these conditions, the current lead is overheated, or the temperature at some position downstream of the room temperature terminal end is higher that T_W (300 K). For values of $R_{HTS\text{-}HX}$ lower than 2 nΩ, the overheating in the TF and PF/CS cases leads even the temperature at the warm end of the meander-flow heat exchanger to exceed T_W. In Fig. 7.7, points at $T_{Cu,W} > 300$ K refer to these cases.

In conclusion of this section, the comparison between the ITER requirements and the results presented in this work is summarized in Tab. 7.7.

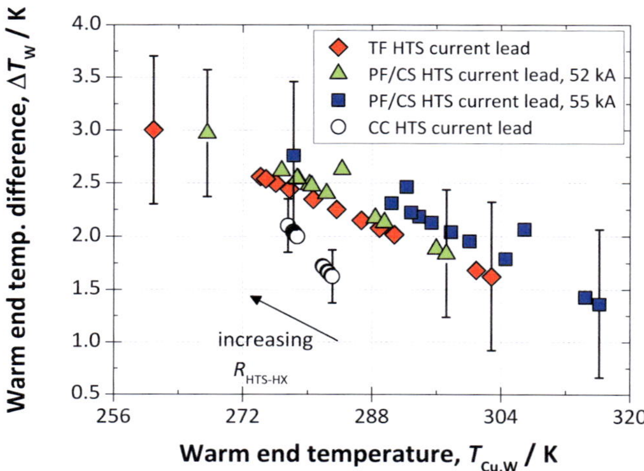

Fig. 7.7: Temperature difference at the warm end of the meander flow heat exchanger as a function of the copper temperature at the same location.

Tab. 7.7: Comparison between ITER requirements and results computed in this paper.

(*) For the case with adiabatic transition between the HTS module and the meander flow HX.

(**) Pressure drop across the HTS current leads and cryogenic feeding system.

	TF		**PF/CS**		**CC**	
	ITER spec.	**1-D model results**	**ITER spec.**	**1-D model results**	**ITER spec.**	**1-D model results**
Maximum He flow rate through the HX, $\dot{m}$ / g/s	4.8	4.878± 0.006(*)	3.85	3.843± 0.005(*)	0.7	0.651± 0.006(*)
Maximum heat load at the cold end, $\dot{Q}_C$ / W	15	13.73	12	11.07	3	1.32
Maximum pressure drop over the HTS current lead / MPa	0.2(**)	0.101± 0.011	0.2(**)	0.072± 0.008	0.2(**)	0.004± 0.0004

7.4 Summary

In this Chapter, the steady state, thermal-hydraulic performance of the ITER HTS current leads have been analysed with a 1-D full length modelling approach.

The design of the HTS current leads has been kept fixed; the design's sensitivity to the variations of the resistance at the warm end of the HTS module and of the cooling condition of the transition between the HTS module and the meander-flow heat exchanger has been analysed. The computed results have been compared with the ITER requirements relevant for the HTS current leads.

According to the results, the design of the ITER HTS current leads fulfils the requirements as far as the maximum allowed helium mass flow rate and the heat leak at the cold end of the current leads are concerned.

It has also been shown that it is not always possible to have $\dot{Q}_W = 0$ for the cooling conditions at the transition between the HTS module and the meander flow heat exchanger and the contact resistances at the warm end of the HTS module covered in this work. In these cases, some external contributions like e.g. heaters or water cooling cartridges are needed to stabilize the temperature profile at the room temperature terminal of the HTS current leads, as currently foreseen in the ITER design.

The pressure drop computed from our analysis across the entire HTS current lead is always significantly lower than the limit given in the ITER specifications. However, a direct comparison is not possible, because the ITER requirements account also for the pressure drop in the cryogenic feeding system. Since the contribution of the cryogenic feeding system to the total pressure drop has not yet been quantified, the computed results have been used instead to define the maximum, allowed pressure loss within the feeding system.

8 Conclusion and perspectives

Current leads transport the electrical current from a room temperature power supply to the superconducting magnets housed into the cryostat of a fusion reactor. They are relevant components because they represent a preferable gate for heat power deposition inside the cryostat.

Current leads can be made of resistive conductors according to a design procedure that is well established and understood (see, for instance [Wils89, p.256]).

A convenient alternative is represented by HTS current leads, which consists of two main components connected in series:

- a normal conductor, which operates in the temperature range $T = 70 - 300$K and requires cooling (room temperature terminal and heat exchanger),
- a HTS module provided with HTS conductors which operates in the temperature range $T = 5 - 70$ K.

Since resistive losses are strongly reduced at the cold end, HTS current leads require a lower cooling power to be operated than equivalent normal resistive current leads.

As for resistive current leads, the design of HTS current leads aims at minimizing the cooling power required to operate the current lead itself. This procedure is referred to as *optimization* of the (HTS) current leads and needs to be performed due to the variety of applications and possibly occurring boundary conditions.

To make the design and the optimization process more accurate and effective, novel techniques for the numerical analysis of HTS current leads have developed in this work.

In the first place, the helium thermal-fluid dynamics inside the resistive heat exchanger has been studied in the so-called meander flow geometry. Heat exchangers characterized by this geometry are mounted in the HTS current leads for the stellarator W7-X and will be used for the HTS current leads of the tokamaks JT-60SA and ITER.

They have also been mounted in the HTS current leads used to power the superconducting magnet system of the Large Hadron Collider at CERN.

Taking advantage of the periodicity of the meander flow geometry and of the limited change in the helium properties on a single period, a Computational thermal-Fluid Dynamic technique based on the periodic modelling has been developed and applied to the systematic analysis of the helium flow.

As a result of the application of the periodic modelling it has been possible to:

- characterize the helium flow and derive formulations for the geometrical quantities that are relevant for the thermal-fluid dynamics in the meander flow geometry, namely the helium flow cross section A_{He} (§ 5.1.1) and the hydraulic diameter d_{h} (§ 5.1.2),
- characterize the helium flow regime and define ranges depending upon the Reynolds number Re for the laminar flow and turbulent flow (§ 5.2.2 and § 5.2.1),
- derive for both flow regimes (i.e. laminar and turbulent) correlations for the pressure drop coefficient ζ and for the Nusselt number Nu, which depend on the Reynolds number Re and other dimensionless ratios of meander flow geometry quantities (§ 5.3 and § 5.4).

With these correlations, the performance of heat exchangers characterized by different meander flow geometry arrangements can be analysed in detail and optimized design solutions can be readily found.

The correlations presented in § 5.3 and § 5.4 have been applied to the first 1-D full-length modelling of all ITER HTS current lead types, i.e. the HTS current leads for the Toroidal Field coils, for the Central Solenoid and Poloidal Field coils and for the Correction coils. The modelling aimed at a predictive analysis of the performance and at an independent verification of the HTS current leads' design (Chapter VII). This analysis has covered the steady-state operation of all HTS current lead types and quantified the influence of the contact resistance as well as of the cooling conditions at the interface between the HTS module and the heat exchanger. The computed results have been then compared to the relevant ITER requirements.

This comparison has shown that the HTS current leads performance predicted with the 1-D modelling generally fulfil the ITER requirements. Variations of the contact resistance and of the cooling conditions at the interface between the HTS module and the heat exchanger slightly influence the required helium mass flow rate and the temperature profile at the warm end of the HTS current leads. Regarding the helium mass flow rate, it is always below the ITER requirements threshold except for a few and very pessimistic cases involving the HTS current leads for the Toroidal Field coils; nevertheless, the difference never exceeds 2% more than the requirements and does not lead to any major issue. Depending on the contact resistance and on the cooling conditions at the interface between the HTS module and the heat exchanger, external heating or cooling systems may be needed to shape the temperature gradient at the warm end of the HTS current leads. The introduction of such mitigation systems is already foreseen in the design. According to the computed results, the maximum power to be delivered for shaping purposes is below 1.2 kW (per HTS current lead, Toroidal Field coil case).

The 1-D modelling has also provided details on the heat load at the cold end of the HTS current leads and the pressure drop occurring in the helium cooling circuit inside them. Regarding the heat load at the cold end, the corresponding ITER requirement is fulfilled. On the other hand, the computed pressure drop cannot be compared directly with the ITER requirement because the latter indicates the maximum allowable pressure over the HTS current leads and the cryo-feeding system. Since the cryo-feeding system is still in the design phase, the computed pressure drop in the HTS current leads has been used to estimate the maximum allowable pressure drop in the cryo-feeding system.

For the design of the HTS module, numerical 2-D axis-symmetric and 3-D reduced models have been developed. The models have been applied to the analysis of the HTS module of the 70 kA ITER Demonstrator and of a W7-X HTS current lead. The computed results have been compared to both steady-state (normal operation) and time-dependent (LOFA accident) sets of experimental results.

It has been shown that both 2-D axis-symmetric and the 3-D reduced model reproduce the steady-state experimental results (§ 6.2.1).

The time-dependent modelling of LOFA accidents has shown that both 2-D axis-symmetric and 3-D reduced model predict a more severe transient: indeed, quench occurs after a time period about 10 - 20% shorter than the experimental one. Although this can be due to the simplifications introduced with the numerical models themselves, it might be more likely related to the accuracy of the material properties implemented therein and in particular to the temperature dependence of the electrical conductivity of the superconductor. More accurate material properties may therefore improve the time-dependent predictions of both the 2-D axis-symmetric and the 3-D reduced model. However, the results show that both models allow a conservative handling of the problem and predict an anticipated quench with respect to an actual HTS current lead (§ 6.2.2).

According to these considerations, both 2-D axis-symmetric and 3-D reduced model provide:

- a reliable prediction of the steady-state operation of the HTS module,
- a conservative description of the evolution during a LOFA accident.

Regarding the choice of the model, no significant differences in the computed outcomes have been noticed among the 2-D axis-symmetric and the 3-D reduced model. Nevertheless, the CPU time required to solve a 2-D axis-symmetric model is considerably shorter (i.e. at least of a factor ten) than for the corresponding 3-D reduced model.

Regarding further developments of the HTS current leads' technology, the main changes deal with:

- cooling the heat exchanger with nitrogen vapour instead of gaseous helium,
- introducing *RE*BCO coated conductors instead of BSCCO.

Cooling the heat exchanger with nitrogen vapour requires liquid nitrogen to be evaporated from a bath at the cold end of the heat exchanger itself. Pressure at this position has to be slightly higher than atmospheric pressure in order to have a sufficient pressure head to maintain the vapour flow through the heat exchanger;

therefore, the temperature at the cold end of the heat exchanger would be higher (i.e. $T > 77$ K, depending on the pressure) than today's HTS current leads. Correlations presented in § 5.3 and § 5.4 are applicable to the nitrogen vapour flow so long buoyancy effects are negligible and the Reynolds number is within the range of applicability. However, validation of the correlations against nitrogen vapour flow experimental data would be worthwhile.

The use of *RE*BCO coated conductors instead of BSCCO introduces some issues in the manufacturing of the HTS module as well as in the detection of an eventual quench (as discussed in Appendix A). The modelling technique discussed in Chapter VI can be in principle adapted to the study of HTS modules with *RE*BCO coated conductors; however, it is worth mentioning that they would aim at modelling a HTS module whose superconducting part consists of *RE*BCO and not at modelling in detail the *RE*BCO coated conductors themselves. Indeed, due to their geometry, a detailed modelling of *RE*BCO coated conductors can only be applied on a much shorter length scale than the one of a HTS module with BSCCO (the length scale difference is in the order of 10^5).

Appendix A HTS conductors of technical interest

In this section, a brief overview on the HTS conductors relevant for HTS current lead applications is presented. For a more detailed analysis on the state of the art of superconducting applications for the nuclear fusion science and technology, with particular emphasis on the transition from LTS superconductors to HTS conductors, the author recommends the introductory part of [Bar13].

Bismuth-strontium-calcium-copper-oxide, BSCCO

The term BSCCO refers to a family of cuprate HTS superconductors containing bismuth, strontium, calcium, copper and oxygen, but no rare-earth element. The general stoichiometric formula is $Bi_2Sr_2Ca_{n-1}Cu_nO_{2n+4+x}$, with n varying in the range $n = 1 - 3$. BSCCO superconductors have a so-called "perovskite" structure, where the superconductivity takes place in a copper oxide plane. The superconducting properties of these materials have first been observed in 1988 [MTF88]. Depending on the index n, the properties of the BSCCO materials vary considerably [Wika]. For this reason, only two compounds are the best suited for technical applications: the BSCCO 2212 ($n = 2$) and the BSCCO 2223 ($n = 3$).

BSCCO 2212

Conductors based on the BSCCO 2212 can be manufactured in forms of wires or tapes. Although the BSCCO 2212 can be also used as bulk material, this form is not suitable for nuclear fusion applications. Therefore only the wire and tape forms will be considered in the following.

Round wires of BSCCO 2212 can be manufactured from a powder containing the necessary reactants and following a variety of possible procedures [MMM03] as the powder-in-tube technique, coating techniques (thin film deposition and thick film deposition) or partial melting processes. Typically, also composite materials as silver/ silver-magnesium are needed. The resulting wires consist of thin BSCCO 2212 filaments embedded inside a matrix of silver/silver-magnesium. Nevertheless, the optimization of the superconducting properties requires complicated heat treatment processes of the wires. The potential benefits of these procedures are affected by some technical difficulties, which lead to the formation of pores inside the wires themselves. Improvements to the heat treatment procedures are needed in order to suppress the formation of pores. Indeed, the pores strongly affect both the electrical and the mechanical properties. Furthermore, the significant fraction of silver needed increases the raw material costs, reduces the mechanical properties and sharps the radioactive activation in case of neutron irradiation. Wires of BSCCO 2212 could be used for large superconducting magnets, but, at present, they do not represent a viable option [Bar13, p. 19].

On the other hand, tapes made of BSCCO 2212 have been manufactured and successfully used in HTS current leads. For instance, for the HTS current leads of the LHC a prototype with dip coated BSCCO 2212 tapes has been proposed in [TMC97], whereas an alternative based on BSCCO 2212 Melt Cast Material in [HKS99]. Alternatively, electronically deposited BSCCO 2212 tapes have been developed for HTS current leads rated at 1000 A in [LDR99]. HTS current leads made of BSCCO 2212 are presently manufactured on a commercial base as well [Nex].

BSCCO 2223

Conductors based on the BSCCO 2223 are manufactured in form of tapes with the powder-in-tube process. In this process, small tubes are filled up with barium, strontium, calcium and copper, which have previously been milled and sintered. Tubes are then closed and extruded. Several tubes are then bundled together and extruded again, in a process that is repeated several times. As it can be seen from the general stoichiometric formula of BSCCO HTS superconductors, oxygen is also needed. For this reason, the BSCCO 2223 tapes have to undergo a heat treatment in an oxygen-rich atmosphere. The aim is to increase the oxygen deposition in the material. From this point of view, it is clear that the material constituting the tubes has to satisfy the requirements in terms of oxygen permeability: tubes are therefore made of silver [RWH92]. Since the current carrying capabilities depend on the grain orientation of

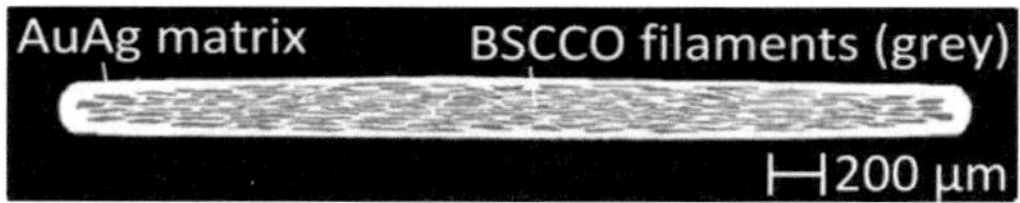

Fig. A.1: Cross section view of a BSCCO 2223 tape. The silver-gold matrix is shown in white and the BSCCO filaments in gray. Source [GSR08].

the BSCCO 2223 material, the bundled tubes are finally rolled to a flat conductor [Kom95, p.92]. A typical cross section of a BSCCO 2223 tape is shown in Fig. A.1.

On average, about half of a BSCCO 2223 tape consists of a silver matrix, which couples the BSCCO 2223 strands and acts as electrical and thermal stabilizer. On the other hand, some not trivial side-effects have to be considered: in the first place, silver largely influences the cost of the tapes [Hul03]; it increases the average heat conductivity of the tapes; it affects the mechanical behaviour of the tapes [Bar13, p. 17] and last, but not least considering the applications, for instance, in nuclear fusion, it is activated by neutron irradiation.

Against this background and considering the fast development of *RE*BCO coated conductors, the production of BSCCO 2223 has reduced over the last decade. Presently, only the company Sumitomo is still producing BSCCO 2223 tapes.

The technical application of the BSCCO 2223 has taken advantage of the superconducting properties of the tapes. The high critical temperature T_c = 108 K (B = 0 T), the steep inverse dependence of the critical current density J_c to the magnetic field as well as the higher sensitivity to the magnetic field perpendicular to the tapes than parallel to it make the BSCCO 2223 tapes suited for applications at low and intermediate magnetic fields ($B < 0.5$ T). Examples are the HTS current leads [BBB12, BMM03, HAA04, HFK11] and the HTS power cables [MYI07].

The BSCCO 2223 tapes for HTS current leads of W7-X and JT-60SA have required a gold-silver matrix [HFK08] instead of pure silver. The goal was to reduce the heat conductivity and therefore the heat conducted towards the cold end of the current leads themselves. Furthermore, the total amount of silver was reduced. This is of great benefit, in particular for W7-X since the HTS current leads are located inside the biological shield [FHK09].

As shown in Chapter III and VI, the BSCCO 2223 tapes have been soldered into stacks and then assembled into panels for the arrangement inside the HTS current leads. Since the electrical, mechanical and thermal properties of the stacks are not trivially derivable from the BSCCO 2223 tape's properties, a detailed characterization of both the tapes and the stacks' properties is required, as shown in [HFK08, KSW09, SWH09] for the case of W7-X HTS current leads.

Recently, the results of the operation of the HTS current leads powering the LHC magnets have been presented [Bal12], showing the reliability and effectiveness of HTS current leads based on BSCCO 2223 tapes.

Rare earth-bismuth-copper-oxide, *RE*BCO

The term *RE*BCO refers to the family of rare-earth-barium-copper-oxide HTS conductors. The general stoichiometric formula is Rare-$Earth_1Ba_2Cu_3O_{7-x}$ and the structure is a defective perovskite crystal [Oak96]. The electrical properties of *RE*BCO conductors strongly depend on the oxygen saturation x: at a low oxygen saturation ($0.55 \leq x \leq 1$) the material behaves as an electrical insulator; the superconducting behaviour is first observed for $x > 0.55$, whereas the maximum critical temperature T_c is achieved if the oxygen saturation is $0 < x \leq 0.2$ [Oak96]. The superconducting state of *RE*BCO conductors is also dependent on the crystalline structure: disruptions or bad oriented grains reduce the current carrying capabilities [GMP04]. The superconducting properties of these materials where discovered in 1987 on compounds using yttrium as rare earth. For this reason, this class of superconductors is also referred to as YBCO HTS conductors instead of *RE*BCO.

*RE*BCO HTS conductors can either be grown as a bulk material, or deposited in form of thin film on metal substrate tapes. In the second case, the *RE*BCO HTS conductors are called coated conductors. For applications in the fusion energy science and technology as well as in the power engineering, *RE*BCO tapes or *RE*BCO coated conductors are very interesting.

The layout of a *RE*BCO coated conductor is qualitatively shown in Fig. A.2. Tapes are available with width varying in the range 4 – 40 mm and thickness in the range 50 – 200 μm [Bar13, p. 20]. The thickest layer of the tape is the so-called substrate

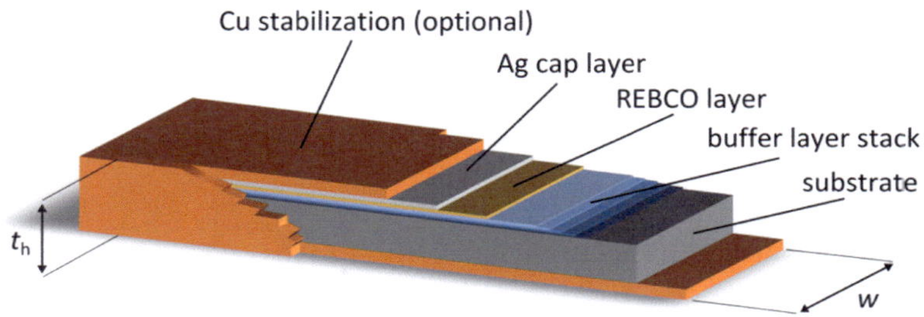

Fig. A.2: Layout of *RE*BCO coated conductors. The layers are scaled arbitrarily. Picture from [Bar13].

layer, whose thickness can vary from 50 to 120 μm. The substrate layer is made of metal, normally Hastelloy®, stainless steel or nickel alloys [Bar13, p. 20]. The substrate layer is coated with several metal-oxide buffer layers. The buffer layers prevent diffusion between the substrate and the *RE*BCO layer and compensate the lattice mismatch, allowing therefore the growth of homogeneous *RE*BCO layers. As mentioned above, the superconducting properties of the *RE*BCO tapes strongly depend on the alignment of grains, which has to be guaranteed over the length of the tape (up to ~1 km [Sel11]). Several techniques can be used to achieve the alignment, as the rolling-assisted-bi-axial-texture (RABITS), the alternating-beam-assisted-deposition (ABAD) o the bi-axial texture in the buffer layers, with ion-beam-assisted-deposition (IBAD) [Bar13, p. 20]. Above the buffer layers, a thin film of rare-earth-barium-copper-oxide is then deposited, with a constant thickness which can vary in the range 1-3 μm. The last thin layer is normally made of silver, silver-gold or gold and it is required for thermal and electrical stabilization [Bar13, p. 20]. The coated conductor in Fig. A.2 is characterized by a further, optional coating made of copper. Its thickness can vary in the range 20 – 100 μm and can be used for further electrical stabilization of the tape.

Nowadays, the research and the development of *RE*BCO tapes are on-going, but a high-quality commercial mass production remains challenging. Nevertheless, there are no doubts about the central role that *RE*BCO tapes will play in the near future of applied superconductivity. Considering the nuclear fusion science and technology, the electrical, thermal and mechanical characteristics make *RE*BCO tapes the most suited

HTS conductors for realizing large superconducting magnet systems [Bar13, p. 27]. Most probably, the *RE*BCO tapes will be used for the next generation of HTS current leads as well, replacing the BSCCO 2223 tapes. However, it is worth mentioning that rather the lacking availability of BSCCO 2223 tapes will trigger the transition to *RE*BCO tapes, not physical or technological limits of the BSCCO 2223.

Moreover, the successful application of *RE*BCO tapes in HTS current leads has to overcome some technical issues: in the first place, *RE*BCO tapes have to be connected to the ends of the HTS module. Joints with low resistance have therefore to be realized for all *RE*BCO tapes on the HTS module. A viable and effective solution has not been found yet. Secondly, in case of quench a rapid detection is compulsory; otherwise, quenched *RE*BCO tapes can be destroyed. An attempt to increase the electrical stability of the tapes would lead to unsatisfactory results. Indeed, the external coating of the tape shown in Fig. A.2 should be thicker, resulting in a – perhaps fast enough – quench detention, but increasing the cross section and the equivalent heat conductivity with resulting higher heat fluxes towards the cold end of the HTS current lead. In this case, benefits of using HTS current leads could be strongly reduced.

Appendix B Forthcoming nuclear fusion experiments

This section gives an overview on the nuclear fusion reactor projects that are relevant for the present work. Two of them are tokamaks, namely the International Thermonuclear Experimental Reactor and the Japanese Torus 60SA; whereas the third is the stellarator Wendelstein 7-X. A fourth machine will be also mentioned, the so-called DEMO fusion reactor. It will not be referred to as an actual machine; indeed, for the time being, several projects have been proposed by agencies throughout the world [Bar13, p. 41]. The name DEMO will rather be used to denote the next generation of nuclear fusion reactor based on magnetic confinement, which is expected to be the precursor of commercial nuclear fusion reactors.

Tokamak

International Thermonuclear Experimental Reactor, ITER

The International Thermonuclear Experimental Reactor ITER is presently the leading fusion experiment worldwide. The project has been developed since 1985, when the former Soviet Union, the USA, the European Union and Japan subscribed at the Geneva Superpower Summit the first agreement for jointly developing fusion energy for peaceful purposes [ITEa]. Later on, the People Republic of China, the Republic of Korea (in 2003) and India (2005) also joined the agreement. The final goal was to develop and build a fusion reactor able to demonstrate the feasibility of commercial fusion energy production. Each member had been asked to contribute to ITER with in-kind contributions, i.e. by providing actual parts/components of the fusion reactor [ITEa]. The ITER reactor is presently being built in Cadarache, France. ITER is rated at 500 MW of fusion power and is designed to achieve a ratio of delivered-to-consumed power Q in the order $Q \geq 10$. It represents an extrapolation of approximately a factor 2 in linear dimension from the largest experiments today and will contain a plasma volume of more than 800 m^3 [Sip04]. In order to have Q larger

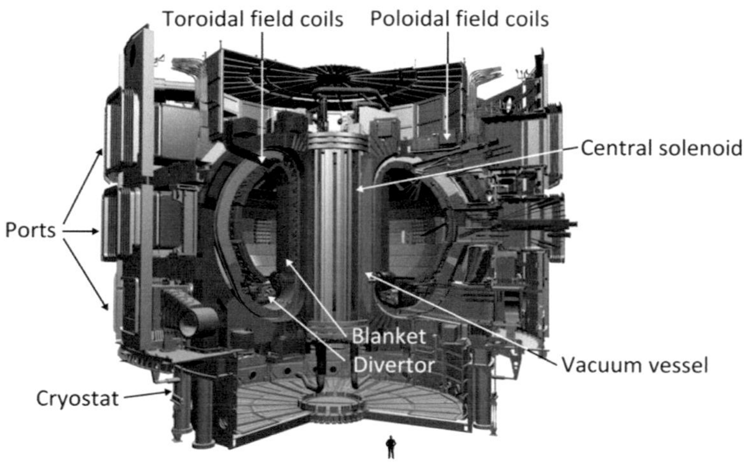

Fig. B.1: View of the ITER tokamak. Source: ITER Organization.

than one, sustained deuterium/tritium plasma predominantly heated by the α-particles produced by the fusion reactions (so-called burning plasma) will be for the first time ignited. Furthermore, the deuterium-tritium will be burn both in inductively driven plasma and in steady-state operation using non-inductive current drive [Sip04].

Besides the plasma physics, the operation of ITER will also address key engineering issues as the plasma facing components, the high heat and neutron fluxes to be handled, the tritium breeding and many more [Sip04]. The solution to these problems will constitute the basics of the development of DEMO.

A view of the ITER reactor with the focus on its main components is shown in Fig. B.1.

The designed superconducting magnet system of ITER consists of 18 toroidal field coils, a central solenoid made of 6 modules, 6 poloidal field coils and 18 correction coils [MDL12]. The toroidal field coils and the central solenoid modules are wound from cable-in-conduit conductors made of Nb_3Sn, whereas the superconductor for the poloidal field coils and the control coils is NbTi. The different choice of superconductor among the coil types depends on the operative magnetic field: since the toroidal field coils and the central solenoid operate at higher magnetic field, Nb_3Sn is used because of the higher critical magnetic field density than NbTi [MDL12]. The cable-in-conduit conductors consist of about 1000 strands of superconductor stabilized with copper cabled around a central cooling spiral tube [MDL12]. The cables for the toroidal field coils are contained in a circular stainless steel jacket; for the other coils, the jacket has a square cross section. The control coils use a reduced size conductor without central channel. All coils are cooled with supercritical helium entering the magnetic system at 4.5 K [MDL12]. The main operative parameters of the ITER coils (toroidal, poloidal field coils and central solenoid) are gathered in Tab. B.1. The superconducting magnet system of ITER will be fed by HTS current leads, as discussed in Chapter VII.

The vacuum vessel is a large torus structure that contains and supports in-vessel components such as the blanket and the divertor. Its main function is to provide a high vacuum for plasma and the primary confinement boundary. It also provides neutron radiation shielding [KNI98]. The vacuum vessel is divided toroidally in 20 sectors. Each sector has a D-shaped cross section approximately 9 m wide and 15 m high. The structure consists of a double wall made of stainless steel shells 40-60 mm thick. The inner and outer shells are joined by welded stiffening ribs [KNI98]. The vessel has 20 vertical, equatorial and diverter ports [KNI98]. Approximately 65% of the volume between the shells is filled with stainless steel plate inserts [KNI98] to provide the required nuclear shielding (with 1-2% boron) [UCE05] or ferromagnetic steelplates to reduce the toroidal field ripple [UCE05]. The vacuum vessel will be maintained at 120 °C by water flowing in two independent cooling loops [KNI98]. The divertor of ITER has represented one key-technology challenge. Indeed, in order to achieve sustained burning plasma operations (300 - 500 s) with $Q \geq 10$, the design had to demonstrate the possibility to control high heat flux transients, to provide a sufficient He pumping and an adequate screening of impurities released as a consequence of intense plasma-surface interactions; last, but not least, the divertor targets must have a tolerable lifetime and a minimized tritium retention [PKL09]. After fifteen years of physics and technology R&D, the design has met the ITER requirements [PKL09]. The diverter consists of 54 fully remotely handleable separate cassette assemblies. Each cassette is 3.5 m long, 2 m high, whereas the thickness increases from the inboard towards the outboard from 0.4 to 0.7 m; the weight is about 9 tonnes. The cassettes are made of a

Tab. B.1: Main parameters of the ITER superconducting magnet system. Source [MDL12].

	Toroidal Field coil	Poloidal Field coil	Central solenoid
Number of coils /modules	18	6	6
Nominal peak field / T	11.8	6	13
Max. operating current / kA	68	45	45
Operating temperature / K	4.5	4.5	4.5
Discharge time constant / s	15	11.5	18
Type of strands	Nb_3Sn	NbTi	Nb_3Sn

cassette body on which four separate plasma facing component units are arranged. The cassette body is made out of stainless steel and, besides providing the structural support for the plasma facing components and a manifolds for their cooling system, it also acts as neutron shielding for the vacuum vessel. Water will be used for the cooling of the divertor cassettes. A large central slot in the cassette body allows for neutrals pumping (provided by a total of eight divertor cryopumps) [PKL09]. Regarding the plasma facing components, the first divertor to be installed for the non-nuclear operational phase (H/He phase) will use carbon fibre composite monoblocks. During the nuclear operation (D/D and D/T) wolfram monoblock targets will be used instead [PKL09].

The blanket as well is a key-system of ITER and one of the most technically challenging for the machine. It accommodates large heat fluxes from the plasma, provides a physical boundary for the plasma transients and contributes to the thermal and nuclear shielding of the vacuum vessel and the external components of ITER [RM11]. The blanket system covers about 600 m^2 and is divided into modules. Each module consists of two major components: a plasma facing first wall and a shield block. The cooling is provided by water at 3 MPa and 70 °C [RM11]. The plasma facing components are shaped to avoid high heat loads in case of panel misalignment and to reduce the eddy current-related loads (they are shaped as "fingers") [RM11]. Two kinds of first wall will be used depending on the heat fluxes, i.e. for heat fluxes up to 5 MW/m^2 or in the order of 1-2 MW/m^2: in the first case, hypervapotron channels made of CuCrZr, whereas in the second stainless steel tubes embedded into

CuCrZr [RM11]. The shield blocks mainly provide nuclear shielding and supply the first wall panels with cooling water.

The ITER auxiliary heating and current drive system consists of neutral beam injection, electron and ion cyclotron resonance heating [Ras11]; the possibility of an upgrade to lower hybrid heating has been foreseen [Ras11]. The neutral beam injection system is rated at 33 MW of power and will be used for electron heating and profile current drive purposes. The electron and ion cyclotron resonance heating are both rated at 20 MW; the first one will provided by gyrotrons at 170 GHz and will be used for electron heating and for profile current drive/control of neoclassical transport modes; the second one will be provided at frequencies in the range 40 - 55 MHz with the purposes of ion and electron heating and central current driving. An upgrade of the heating system with lower hybrid heating at 20 MW will serve for fast electron heating and edge current drive [Ras11].

Japanese Torus 60SA, JT-60SA

The Japanese Torus 60 SA is an experimental satellite tokamak jointly funded from the European Union and the Japan in the frame of the Broader Approach agreement [EU07]. The mission of the JT-60SA is to support the ITER experiment in resolving key physics and engineering issues towards the development of DEMO [JT60]. The device can operate in single or double null configuration and is expected to confine break-even equivalent plasmas for duration longer then the timescales characteristics of plasma processes, to pursue full non-inductive steady-state operation and to establish high-density plasma regimes [IBK11]. The JT-60SA is being constructed at the Naka site of the Japan Atomic Energy Agency and the first plasma is expected in 2016 [IBK10]. An bird-eye view of the JT-60SA is shown in Fig. B.2.

The superconducting magnet system of the JT-60SA consists of the 18 toroidal field coils, a central solenoid with four modules and six equilibrium coils [IBK10, YTK10]. The equilibrium coils are the equivalent of the PF coils in the ITER reactor. The toroidal field coils have a D-shaped form and are wound from a rectangular steel-jacketing NbTi cable-in-conduit conductor [IBK10, YTK10]. The central solenoid is divided in four independent winding pack-modules made of rectangular steel-jacketing

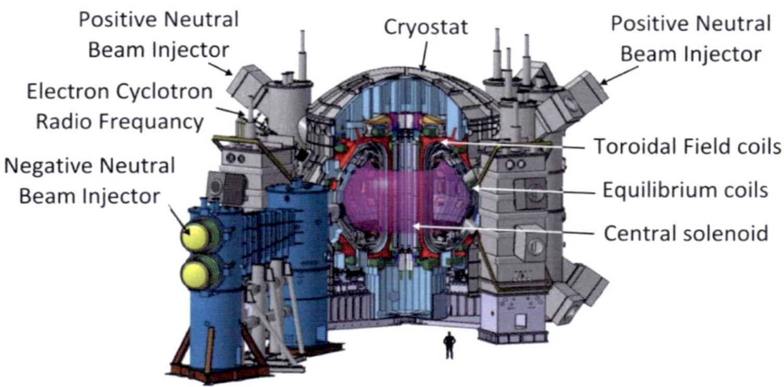

Fig. B.2: Bird's eye view of the JT-60SA tokamak. Published with permission of JT-60SA.

Nb_3Sn cable-in-conduit conductor [IBK10, YTK10].

As the toroidal field coils, also the six equilibrium coils are wound from rectangular steel-jacketing NbTi cable-in-conduit conductor. All superconductors will be cooled with supercritical helium with a coil inlet temperature of 4.5 K [YTK10]. The main operating parameters of the JT-60SA coils are summarized in Tab. B.2.

Beside the superconducting coils, the JT-60SA will also be equipped with three sets of copper coils which are classified as in-vessel components [IBK10]: a pair of fast plasma position control coils, 18 error field correction coils and 18 resistive wall mode control coils.

The superconducting magnet system will be powered with two sets of HTS current leads [FHK09]: 6 HTS current leads rated at 26 kA for the toroidal field coils and 20 rated at 20 kA for the equilibrium coils and central solenoid modules.

The vacuum vessel is composed of 18 toroidal sectors constructed out of SS316L with low cobalt content (Co $<$ 0.05 wt%) [IBK10]. A seismic analysis has been conducted and the project of the supports of the vacuum vessel has been developed in order to withstand accelerations due to a earthquake in the order of 1G. A flow of boric acid water within the vacuum vessel will be used for neutron shielding purposes [IBK10].

The divertor of the JT-60SA should withstand heat fluxes up to 15 MW/m^2 [IBK10]

Tab. B.2: Main parameters of the JT-60SA superconducting magnet system. Source [IBK10].

	Tor. Field coil	**Equilibrium coil**	**Central solenoid**
		EF 3,4 – EF 1, 2 , 5, 6	
Number of coils / modules	18	6	4
Nominal peak field / T	5.65	6.2 – 4.8	8.9
Operating current / kA	25.7	20	20
Operating temperature / K	4.9	5.0 – 4.8	5.1
Discharge time constant / s	10	6	6
Type of strands	NbTi	Nb_3Sn	NbTi
Sup. Strands / copper strands	324 / 162	450 / 0 – 216 / 108	216 / 108

and consists of the inner and outer, V-shaped vertical targets, the private flux region dome and the divertor cassette body. The divertor cassettes are designed to be compatible with remote handling maintenance and allow therefore long-pulse high performance plasma operation with a large neutron yield [IBK10]. In the first phase of tokamak operation, CFC targets (monoblock and bolted target) will be used for the cassettes, whereas tungsten coated CFC monoblocks are left as a future option. All plasma-facing components will be cooled by water at 40 °C [IBK10].

According to [IBK10], the cryostat consists of a body vessel and a base used for the gravity and seismic support of the machine. The vessel body will be made of a single-wall stainless steel shell (SS 304 with low cobalt content Co < 0.05 wt%) designed for a normal operation pressure of 1e-03 Pa (external pressure of 0.1 MPa) and an absolute internal pressure of 0.12 MPa, assuming the loss of helium and water form cryogenic and coolant lines.

The JT-60SA will use a Electron Cyclotron Radio Frequency system for heating the plasma, for current drive purposes and for other relevant plasma operations [IBK11]. In the initial phase of the operations (see below for a rough schedule of the JT-60SA operations), 3 MW of radio frequency power at 110 GHz will be injected into the plasma by 4 gyrotrons of 1 MW each. For a later research phase, named integrated research phase, five additional 110 GHz gyrotrons and power supply sets will be fabricated and installed. In total, 7 MW at 110 GHz will be injected to the plasma by 9 gyrotrons of 1 MW. In all cases, the transmission efficiency is assumed to be 0.75-0.8 [IBK11].

The Neutral Beam Injection system will consist of 12 positive-ion-based units (for perpendicular, counter-tangential and co-tangential beams) with a maximum power of 24 MW [IBK10] and one negative-ion-based unit [IBK11] with a power of 10 MW [IBK10].

The operation schedule of the JT-60SA has been divided into three phases: the initial research phase, the integrated research phase and the extended research phase [JT60].

The initial research phase is further split in two sub-phases: the hydrogen phase and the deuterium phase. In the hydrogen phase the entire system will be commissioned with and without plasma operation; it is expected to last 1-2 years and it will prepare the deuterium phase. The deuterium phase will last 2-3 years. During this phase the remaining commissioning related to neutron production, nuclear heating and radiation safety will be carried out. Furthermore, operational boundaries and experimental flexibilities will be characterized and the target regimes of the JT-60SA have to be studied using short pulse discharges.

The integrated research phase will investigate and demonstrate the main mission of the JT-60SA with high-power long-pulse discharges. As for the initial phase, there will be two sub-phases (I and II) in the integrated research phase as well. In the sub-phase I, the neutron production will be limited in order to allow human access inside the vacuum vessel. Indeed, the commissioning of the remote handling system must be completed during this phase. In the sub-phase II the neutron production limit will be increased and the remote maintenance of in-vessel components will be required. The integrated research phase will probably last more than 5 years.

In the extended research phase the JT-60SA will be operated at higher heating power with a double null configuration. Depending on the progress of tokamak research worldwide different types of diverter targets and first walls will be installed. This last phase is expected to last more than 5 years.

Stellarator

Wendelstein 7-X, W7-X

The Wendelstein 7-X is an experimental stellarator that aims at exploring and demonstrating the reactor potential of the stellarator principle [Wan00]. The total investment costs are jointly carried by the European Union, the Federal Government of Germany and the State of Mecklenburg-Vorpommern [Wan00]. It is presently being built at the Greifswald branch of the IPP (Mecklenburg-Vorpommern, Germany). Start of the operations is scheduled for 2014 [W7X].

The design of the W7-X has been developed to address and study key plasma physics issues and engineering aspects related to the use of superconducting coils, to a modular design of the magnet system and steady-state operation of all components in the stellarator concept. Nevertheless, the W7-X project does not aim at studying a burning plasma; therefore no tritium will be used [Wan00]. A perspective view of the W7-X stellarator and its main components is shown in Fig. B.3.

The W7-X is a drift-optimized stellarator [GWB12] characterized by a helical plasma column with a cross section which varies periodically between a bean shape and a triangular shape. The periodical variation results in a five-fold symmetry of the reactor helical torus. The necessary twist of the magnetic field lines is provided by superimposing a poloidal magnetic field on the main toroidal field [Wan00]. In total, 50 3-D, superconducting non planar coils are needed to generate the magnetic field. The coils are made out of cable-in-conduit conductor, CICC, using NbTi strands. Each CICC consists of 243 copper stabilized NbTi strands with a diameter of 0.57 mm. The strands are enclosed in an aluminum alloy (AlMgSi0.5) jacket (outer dimension 16mm x 16 mm). This aluminum alloy is soft enough to allow the bending of the CICC during the coil production and stable enough to withstand the mechanical loads during the operations. The winding of the non-planar coils consist of 108 turns divided into six double layers. The double layers are connected electrically in series with interlayer joints whose resistance has to be limited at 1 nW. Hydraulically, the double layers are connected in parallel, in a way which secures that the innermost layer (high field layer)

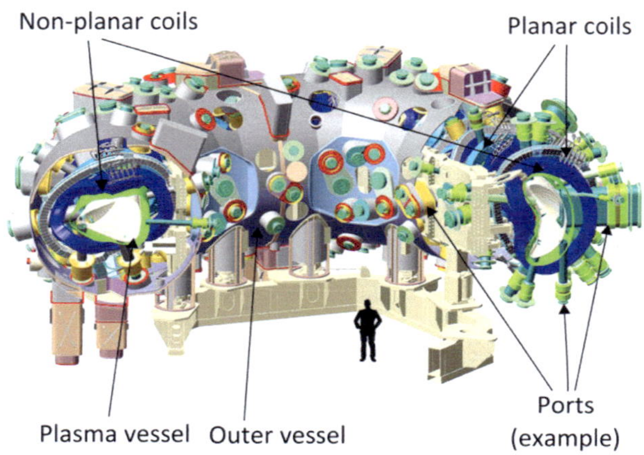

Fig. B.3: View of the W7-X stellarator. Source: Max Planck Institute for Plasma Physics.

always gets the fresh helium from the inlet. The coils' case consists of stainless steel. The manufacturing of the non-planar coils underwent some issues related to standard fabrication processes like welding and insulating; on the other hand, the production and reproducibility of three dimensional coils have been successfully demonstrated. This information as well as a more detailed treatment can be found in [RRE11]. An example of non-planar coil is showed in Fig B.4a.

Besides the non-planar coils, the W7-X superconducting magnetic system has also 20 planar coils. These coils are used to change the magnetic configuration of the machine. The planar coils are assembled over the non-planar coils, at an angle of 20° over the main vertical axis. As superconductor, the same copper stabilized, NbTi CICCs are used. In this case, the winding pack is made from three double layers with 12 turns each electrically connected in series via two interlayer joints with resistance lower than 1 nΩ. Hydraulically, the double layers are connected in parallel similarly to the non-planar coils [RRE11]. An example of planar coil is showed in Fig. B.4b. The arrangement of both non-planar and planar coil in a module of the W7-X stellarator is shown in Fig. B.4c.

The superconducting system of W7-X will be energized by 14 HTS current leads provided by KIT [FHK09]. The main design parameters of both the non-planar and planar coils are gathered in Tab. B.4.

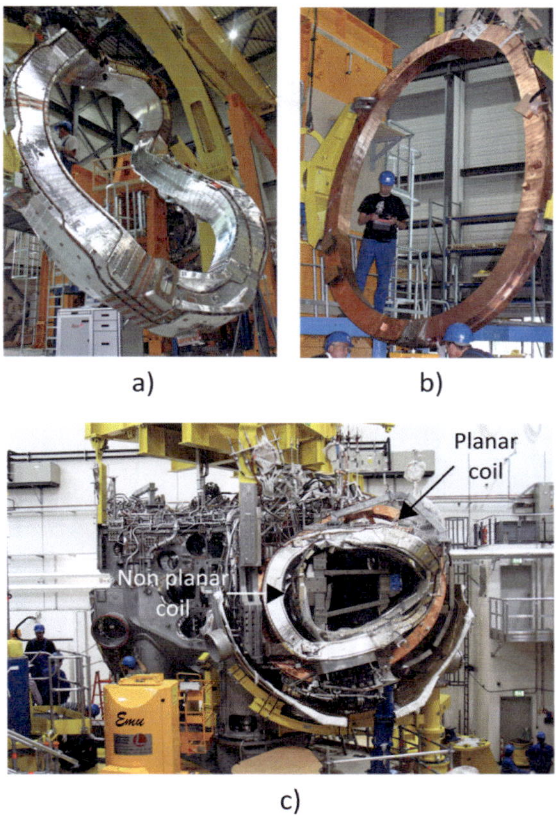

Fig. B.4: a) Superconducting non-planar coil of the W7-X magnet system;

b) Superconducting planar coil of the W7-X magnet system;

c) Assembling of one module of the W7-X; in the foreground the position of a non-planar and of a planar coil can be appreciated.

Tab. B.4: Main parameters of the W7-X superconducting magnet system. Source [RRE11].

	Non planar coil	**Planar coil**
Number of coils	50	20
Nominal peak field / T	11.8	6
Operating current / kA	17.6	17.6
Operating temperature / K	4.5	4.5
Discharge time constant / s	15	11.5
Type of strands	NbTi	NbTi

The cryostat of the W7-X is principally made of the plasma vessel, the outer vessel, the ports and the thermal insulation [WEF03]. The manufacture of the plasma vessel has represented one major challenge due to its shape and the necessity of optimization to give maximum space to the plasma while keeping the necessary clearance against the cold coils. It is made from steel rings bent precisely to the required shape within local tolerances of 3 mm [WEF03]. The cryostat is characterized by the large number of openings for ports (45 for each of the 5 sectors of the reactor), manholes and feed-through. In the plasma vessel the openings are cut by a water jet technique.

The thermal insulation of the cryostat is achieved with high vacuum and several layers of reflecting foils; a further improvement is provided metallic shields, which cover all areas at ambient temperature [WEF03].

The in-vessel surfaces can be classified depending on the heat load they have to withstand. The divertor target plates are hit predominately by hot particles from the plasma and have to withstand heat loads of up to 10 MW/m^2; the baffles, which influence the fluxes and density of neutralized particles in front of the target plates, need to withstand heat loads of 0.5 MW/m^2; the wall protection of the plasma vessel has to withstand heat loads up to 0.2 MW/m^2 [WEF03]. To control the reflux of impurities to the plasma and to minimize the radiation losses all the plasma-facing surface are coated with low-Z material. Furthermore, considering the foreseen operative condition of the W7-X, the plasma-facing components have to be designed for steady-state operation. The divertor target plates cover a surface of 30 m^2. Each target element is composed of a water-cooled metallic support and a flat CFC title. Baffles are installed in front of the target plates and span over the same surface. Baffles are made of graphite tiles clamped to water-cooled support structures. The wall of the plasma vessel covers a surface of area circa 120 m^2; locations where the distance between the plasma boundary and the vessel is small (about 50 m^2) will be covered with clamped tiles as for the baffles. For the remaining surface (about 70 m^2), panels with integrated cooling and coated with B4C will be used [WEF03]. To remove a maximum heating power of 15 MW/m^2 from the divertor and the wall, a water flow of 2750 m^3/h is required. The pressure in the water cycle will be kept at 10 bar in order to avoid boiling [WEF03].

The plasma heating systems are electron and ion cyclotron resonance heating and neutral beam injection. The electron cyclotron resonance heating will deliver 10 MW steady-state heating at 140 GHz with 10 gyrotrons rated at 1 MW each. The ion cyclotron resonance heating provides 2 x 2 MW at frequencies ranging between 25 and 76 MHz. The neutral beam injection system will heat the plasma bulk with a beam power of 5 MW for 10 s using 60 keV deuterium injections. The system can be upgraded to a power up to 20 MW for 15 s [WEF03].

Appendix C Material properties

Helium

The helium properties used for the periodic modeling of the meander flow heat exchanger (Chapter IV and V) have been calculated with MATLAB® routines based on Cryosoft®. The original source for the Cryosoft® data is HEPACK®.

Raw data of the helium density ρ_{He} (Fig. C.1), molecular viscosity μ_{He} (Fig. C.2), specific heat capacity $c_{p,He}$ (Fig. C.3) and heat conductivity λ_{He} (Fig. C.4) are interpolated with respect to the helium temperature T_{He} and pressure p_{He}. For the present work, the range of interest are $T_{He} = 50 - 300$ K and $p_{He} = 0.2 - 0.5$ MPa.

The helium properties implemented in the code CURLEAD (Chapter VII, analysis of the ITER HTS current leads) have been derived from the program HEPROP [Han79].

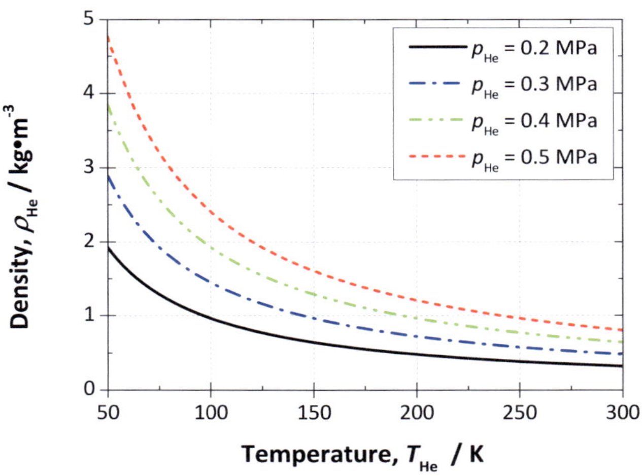

Fig. C.1: Helium density, ρ_{He}. Source Cryosoft®.

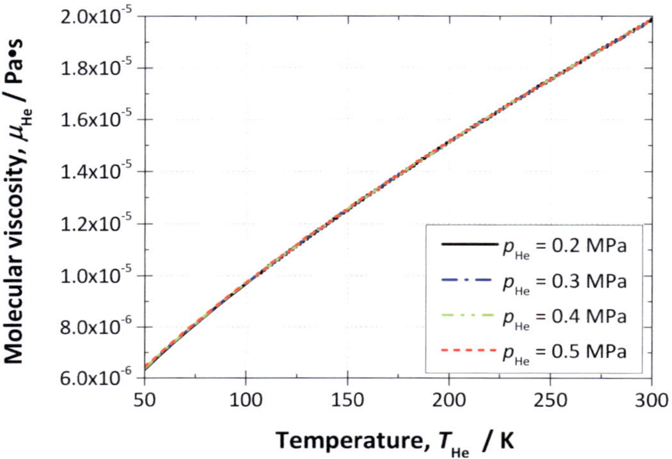

Fig. C.2: Helium molecular viscosity, μ_{He}. Source Cryosoft®.

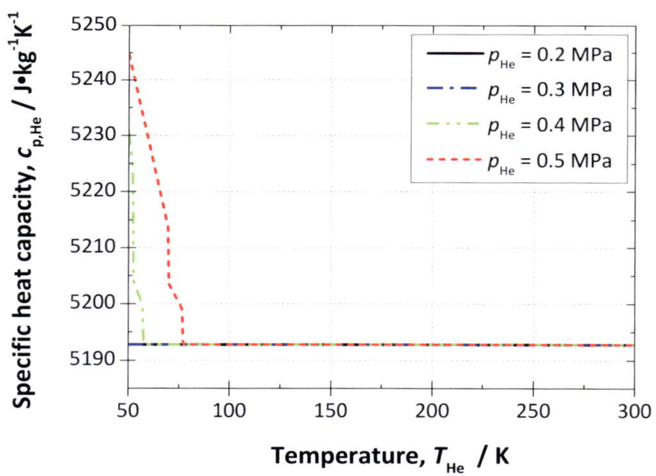

Fig. C.3: Helium specific heat capacity, $c_{p,He}$. Source Cryosoft®.

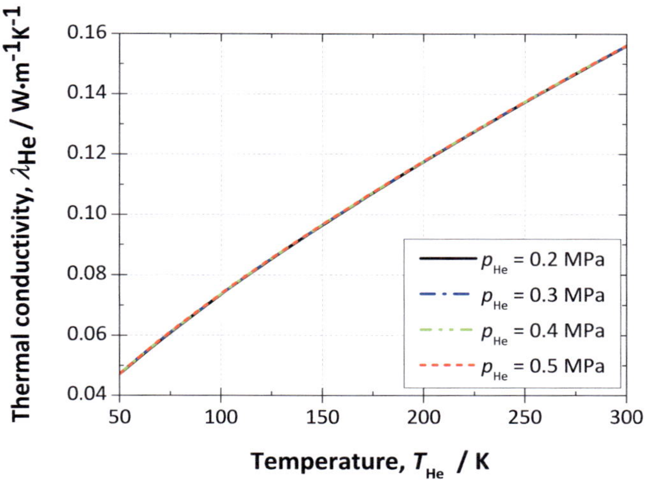

Fig. C.4: Helium thermal conductivity, λ_{He}. Source Cryosoft®.

Copper

For the copper density ρ_{Cu}, the constant value $\rho_{Cu} = 8700$ kg/m^3 has been used.

The values of the copper specific heat capacity $c_{p,Cu}$ are interpolated with respect to the temperature T_{Cu}. References for the raw data are as follows: $T_{Cu} = 0$ - 10 K from [Fic72], $T_{Cu} = 10$ - 400 K from [Joh61], whereas for $T_{Cu} > 400$ K is approximated with Dulong and Petit law [Kit96, p. 127]. The result is shown in Fig. C.5.

The values of the copper thermal conductivity λ_{Cu} are interpolated with respect to the temperature T_{Cu} and the residual resistivity ratio RRR. The reference for the raw data is [Hel13b].

The values of the copper electrical conductivity σ_{Cu} are calculated with the results of the Bloch-Grüneisen law [Czy04, p. 212], as in CURLEAD [Hel89].

The specific heat capacity $c_{p,Cu}$ and the thermal conductivity λ_{Cu} have been plotted as a function of the temperature in Fig. C.5 and Fig. C.6, respectively.

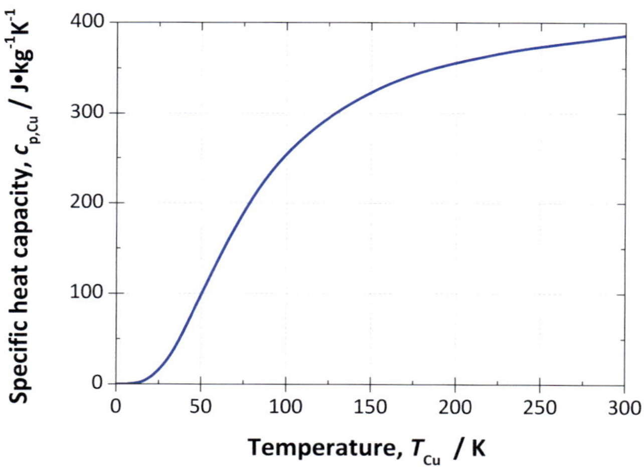

Fig. C.5: Copper specific heat capacity, $c_{p,Cu}$. Source [Fic72, Joh61, Kit96].

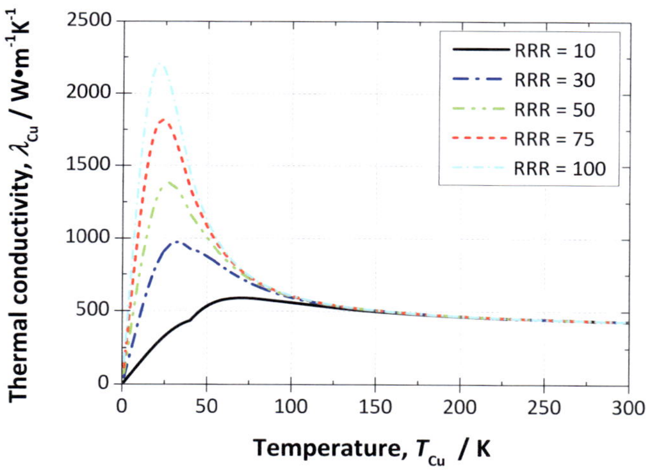

Fig. C.6: Thermal conductivity, λ_{Cu}. Source [Hel13].

Stainless steel

For the stainless steel density ρ_{ss}, the constant value $\rho_{ss} = 7850$ kg/m^3 has been used.

The values of the stainless steel specific heat capacity $c_{p,ss}$ are obtained with the polynomial proposed in [Cryo] and shown in Eq. C.1.

$$\begin{aligned} c_{p,ss} = 10\wedge \Big(&22.006 - 127.553 \cdot \mathrm{Log}(T_{ss}) + 303.647 \cdot \left(\mathrm{Log}(T_{ss})\right)^2 + \\ &-381.01 \cdot \left(\mathrm{Log}(T_{ss})\right)^3 + +247.033 \cdot \left(\mathrm{Log}(T_{ss})\right)^4 + \\ &-112.921 \cdot \left(\mathrm{Log}(T_{ss})\right)^5 + 24.759 \cdot \left(\mathrm{Log}(T_{ss})\right)^6 + \\ &-2.239 \cdot \left(\mathrm{Log}(T_{ss})\right)^7\Big). \end{aligned} \tag{C.1}$$

The values of the stainless steel thermal conductivity λ_{ss} are interpolated with respect to the temperature T_{ss} over the range $T_{ss} = 1 - 400$ K. The resulting polynomial is shown in Eq. C.2. The raw data are listed in [HC77].

$$\begin{aligned} \lambda_{ss} = &-(1.167e-20) \cdot T_{ss}^9 + (2.752e-17) \cdot T_{ss}^8 - (2.698e-14) \cdot T_{ss}^7 + \\ &+(1.436e-11) \cdot T_{ss}^6 - (4.523e-09) \cdot T_{ss}^5 + (8.552e-07) \cdot T_{ss}^4 + \\ &+(9.252e-05) \cdot T_{ss}^3 + (4.680e-03) \cdot T_{ss}^2 + (3.214e-02) \cdot T_{ss} \\ &+(8.597e-02). \end{aligned} \tag{C.2}$$

The values of the stainless steel electrical conductivity σ_{ss} are interpolated with respect to the temperature T_{ss} over the range $T_{ss} = 1 - 400$ K. The resulting polynomial is shown in Eq. C.3. The raw data are listed in [HC77]:

$$\begin{aligned} \sigma_{ss} = &+(1.585e-13) \cdot T_{ss}^8 - (2.776e-10) \cdot T_{ss}^7 + (1.997e-07) \cdot T_{ss}^6 + \\ &-(7.530e-05) \cdot T_{ss}^5 + (1.557e-02) \cdot T_{ss}^4 - (1.610e00) \cdot T_{ss}^3 + \\ &+4.882e01 \cdot T_{ss}^2 + 4.560e01 \cdot T_{ss}^2 + 2.059e06. \end{aligned} \tag{C.3}$$

BSCCO stacks

For the BSCCO density ρ_{BSCCO} the constant value $\rho_{BSCCO} = 7850$ kg/m^3 has been used.

For the BSCCO specific heat capacity $c_{p,BSCCO}$, the function of the temperature T_{BSCCO} in Eq. C.4 has been used (it is assumed that the specific heat of the BSCCO stacks corresponds to that of the silver contained therein; the same assumption has been made in CURLEAD [Hel89]).

$$c_{p,BSCCO} = 10^\wedge\left(-2.7 + 3.6 \cdot \text{Log}(T_{BSCCO}) - 0.41 \cdot \left(\text{Log}(T_{BSCCO})\right)^2 - 0.064 \cdot \left(\text{Log}(T_{BSCCO})\right)^3\right). \tag{C.4}$$

For the 6-tape stack used for the W7-X HTS current leads, the BSCCO heat conductivity λ_{BSSCO} has been derived by interpolating the experimental results presented in [HFK08] and shown in Fig. C.7 with respect to the temperature T_{BSCCO}.

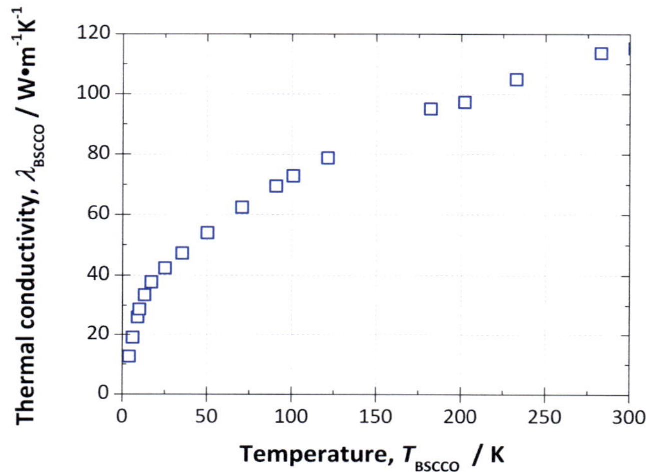

Fig. C.7: Thermal conductivity for the 6-tape stack, λ_{BSCCO}. Source [HFK08].

For the 13-tape stack used for the ITER demonstrator HTS current leads, the heat conductivity is calculated as a power-function of the temperature T_{BSCCO}, as it has been used in [Sch09, pag. 86]. The power-function is reported in Eq. C.5.

$$\lambda_{BSCCO} = -28.214 + 19.460 \cdot (T_{BSCCO}^{0.395}). \tag{C.5}$$

The dependence of the BSCCO electrical conductivity σ_{BSCCO} on the temperature (for $T_{BSCCO} > T_{cs}$) has been described as in [Hel09]:

$$\sigma_{\mathrm{BSCCO}}(T_{\mathrm{BSCCO}}) \sim \left(\frac{T_{\mathrm{BSCCO}} - T_{\mathrm{cs}}}{T_c - T_{\mathrm{cs}}}\right)^{-m}, \tag{C.6}$$

where T_{cs} = 85.8 K for the HTS module of W7-X and T_{cs} = 78 K for the HTS module of the 70 kA Demonstrator; whereas, T_{c} = 95.5 K or the HTS module of W7-X and T_{c} = 91.3 K for the HTS module of the 70 kA Demonstrator [Hel09, HDF11b]. In more detail, values for the W7-X case have been experimentally derived as explained in [HDF11b, p. 65 - 68]. The exponent m is $m = 11$.

Appendix D Grid independence analysis

CtFD periodic modelling

As explained in the fourth Chapter, the mesh used to discretize the computational domain consists of several layers of prism cells in the boundary layer region and of polyhedral cells in the core mesh. An overview on an exemplary grid is shown in Fig. D.1.

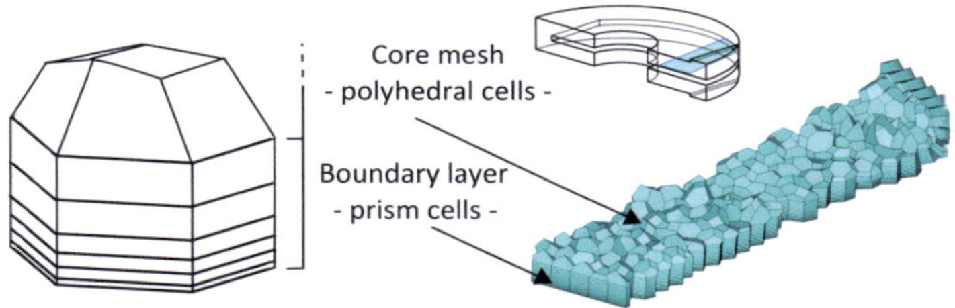

Fig. D.1: Meshing technique for the periodic model computational domain.

The grid independence analysis has investigated the dependence of the computed pressure drop Δp_{He} on the size of the polyhedral cells as well as on the size and the number of layers in the boundary layer. This procedure has been applied to grids for the turbulent regime and for the laminar regime.

Turbulent regime grid analysis

The polyhedral cells constituting the core mesh are built using the mesh generator in Star-CCM+ with, on average, 14 faces [StaCC]. The characteristic dimensions of the polyhedral cells range from 0.5 mm to 0.9 mm; the variation of the dimension mainly depends on the fin distance t, the cut-off c_0 and the position inside the computational domain.

The mesh in the boundary layer has to fulfill the requirements of the adopted turbulence model in terms of height of the first layer departing from the wall (constraint on the *y+* parameter, see [Sta08]), number of layers and prism height increment moving from the first layer towards the core flow region. For the present CtFD analysis, the boundary layer grid has the following characteristics:

- the height of the first layer is homogenous all over the computational domain and has been chosen to have y+ ~ 1, in agreement with the requirements of the Low *Re k-ω* SST model in Star-CD [Sta08],
- in general ~ 15 prism layers have been used in the boundary layer,
- the height of each layer grows by a factor 1.1 with respect to the previous one.

Meshes generated with this scheme have been proven against grid independence on some test meander flow geometry models: grids with different number of layers in the boundary layer and dimension of the polyhedral cells have been tested, showing the solution to have a high degree of grid independence when the parameters are as described above. The plot D.2 shows an example for the grid independence analysis.

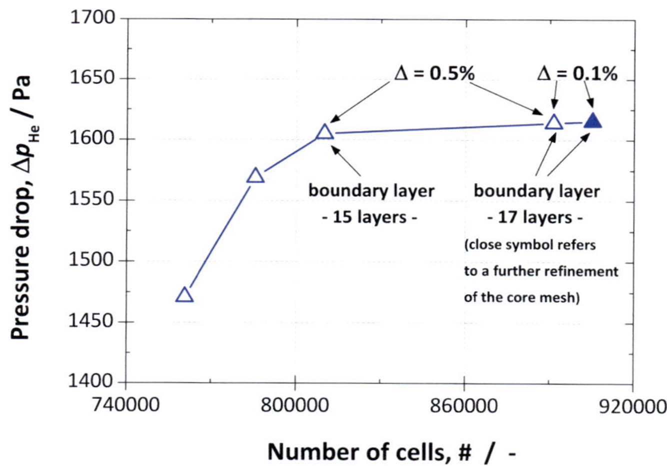

Fig. D.2: Grid independence analysis on a mesh for the turbulent regime analysis. The pressure drop has been computed refining the computational grids both in the core mesh and the boundary layer. Departing from ~800000 cells (15 layers in the boundary layer and core mesh with dimension 0.5-0.9 mm), further refinements of the grid does not lead to any significant change in the computed result.

Laminar regime grid analysis

Typically, the numerical solution of a laminar flow problem does not require a particularly fine mesh close to the boundary. Nevertheless, as stated in [StaCC], it is recommended to provide a regular grid close to the wall boundaries. For this reason, the boundary layer consists of homogeneously distributed prisms cells (with a coarser distribution with respect to grids for the turbulent regime): at least two layers of prisms cells have been used. The polyhedral cells have on average 14 faces [StaCC] and their maximum size varies in the range 0.5-0.9 mm, mainly depending on the fin distance t and on the location inside the computational domain. Computational results obtained on these grids been proved against grid-independence.

An example of grid independence study for the first meander flow geometry in Tab. 4.3 is shown in Fig. D.3. Open symbols refer to grids having the boundary layers made of two layers, but with increasing refinements of the core mesh. This scheme produces mesh with a satisfactory independence on the grid size above 200 kcells (i.e. when maximum size of the polyhedral cells varies in the range 0.5-0.9 mm).

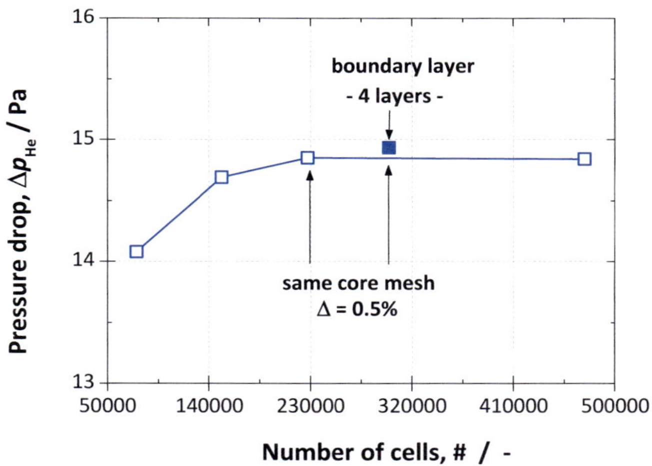

Fig. D.3: Grid independence analysis on a mesh for the laminar regime analysis. The pressure drop has been computed refining the computational grids in the core mesh. Departing from ~200000 cells (2 layers in the boundary layer and core mesh with dimension 0.5-0.9 mm), further refinements of the core mesh do not lead to any significant change in the computed result. For the sake of completeness, the boundary layer of the grid with 230000 cells has been refined and the pressure drop computed (close symbol). In this case as well, the refinement does not lead to any significant change in the computed result.

The close symbol refers to the refinement of the boundary layer when the core mesh is grid independent.

As it is indicated, doubling the number of layers in the boundary layer results in about 0.5% larger pressure drop.

HTS module modelling

The grid independence analysis for the modelling of the HTS module has been performed on a 2-D axis symmetric problem. The results of the analysis have been then adopted also for the generation of the three-dimensional meshes. The grid independence analysis has investigated four mesh configurations as well as the effects on the computed results of their refinement.

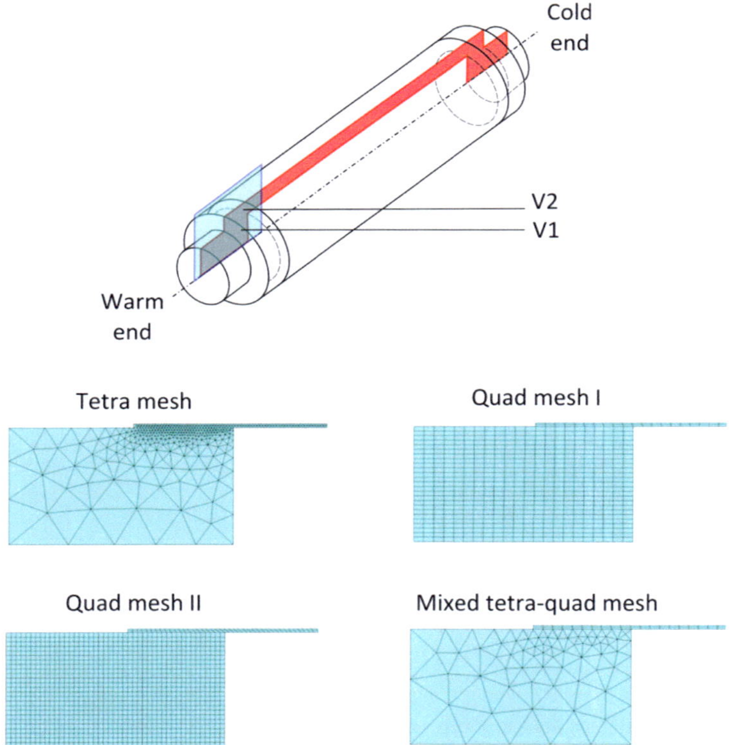

Fig. D.4: Section of the HTS module which represents the 2-D axis-symmetric computational domain and mesh configurations tested in the grid independence analysis..

As shown in Fig. D.4, the four mesh configurations are:

- Tetrahedral, or tetra mesh (triangles),
- Quad mesh I (rectangles),
- Quad mesh II (squares), and
- Mixed tetra-quad mesh (triangles-rectangles).

The configurations Quad mesh I and Quad mesh II have node distributions with different aspect ratios along the computational domain. The configuration Mixed tetra-quad mesh takes advantage of a quad mesh (type I) for discretizing the slender parts of the computational domain, whereas a typical tetrahedral mesh is used elsewhere.

Computed results have been analysed for increasing mesh refinements. As parameter for the comparison, the voltage drop ΔV at the warm end transition normal-super conductor has been used. The voltage drop is calculated as $\Delta V = V2 - V1$. The locations of the voltage sensors $V1$ and $V2$ can be seen in Fig D.4. The results of the grid independence analysis are gathered in Fig. D.5. Results computed on grids with more than 20 kilo-knots are very close, despite the mesh configuration. It can also be see that meshes containing a number of nodes higher than 2e04 do not lead to any relevant change in the computed results.

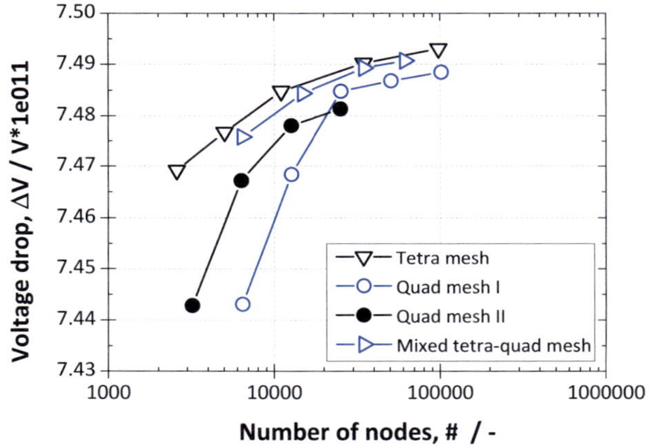

Fig. D.5: Grid independence analysis on four mesh configurations. The voltage drop V2-V1 (see Fig. D.4) has been computed on refined computational grid of each configuration type. Results are can be considered grid independent if computed on grid with more than 20 knots, despite the mesh configuration.

Table of symbols

Latin

A	Surface area / cross section / m^2
a	Width of the helium cross section / m
c_o	Cut-off / m
c_p	Specific heat capacity / $J/kg \cdot K^{-1}$
d_h	Hydraulic diameter / m
d_i	Central bar diameter / m
d_o	Outer fin diameter / m
h	Heat transfer coefficient / $W/m^2 \cdot K^{-1}$
I, $\boldsymbol{I}$	Electrical current / A
J, $\boldsymbol{J}$	Electrical current density / A/m^2
l, L	Length / m
L_o	Lorentz number / $W \cdot \Omega / K$
$\dot{m}$	Mass flow rate / kg/s or g/s
Nu	Nusselt number / -
p	Pressure / Pa
P	Cooling perimeter / m
P_t	Total cooling power / W
$\dot{Q}$	Heat load / W
Re	Reynolds number / -
Re^*	Dimensional Reynolds number / m
s	Fin thickness / m
T	Temperature / K
t	Fin distance / m (Chapters II-V and VII); Time / s (Chapter VI)
v, $\boldsymbol{v}$	Velocity / m/s

V	Voltage / V

Greek

η_c	Carnot efficiency / -
β	Expansion factor / -
ε	Turbulent dissipation / m^2/s^3
κ	Turbulent kinetic energy / m^2/s^2
λ	Heat conductivity / $W/m \cdot K^{-1}$
μ	Molecular viscosity / $Pa \cdot s$
μ_t	Turbulent molecular viscosity / $Pa \cdot s$
ρ	Density / kg/m^3
υ	Kinetic viscosity / m^2/s
υ_t	Turbulent kinetic viscosity / m^2/s
ω	Dissipation rate / 1/s
ζ	Pressure drop coefficient / -

Table of abbreviations and indexes

0	Index for the helium flow cross section at the inlet of the fins region (see Chapter V)
b	bulk
BSCCO	Bismuth-strontium-calcium-copper-oxide
C	Cold
CC	Control coil
co	cut-off
cond	conductive
CPU (time)	*Literal translation:* Central Processor Unit. Here used to indicate the period, which is required to solve numerically a model
CS	Central solenoid
Cu	Copper
He	Helium
HTS	High Temperature Superconductor
HX	Heat eXchanger
in	inlet
ITER	International Thermonuclear Experimental Reactor
JT-60SA	JapaneseThorus – 60 Super Advanced
loss	(resistive) losses
LT	Low Temperature
LTS	Low Temperature Superconductor
MF	Meander Flow
o	optimized
out	outlet
PF	Poloidal Field
RT	Room Temperature
ss	stainless steel
TF	Toroidal Field

W	Warm
w	wall
W7-X	Wendelstein 7-X

Literature

[Bal04] BALLARINO, A: *Conduction-cooled 60 A resistive current leads for the LHC dipole correctors*. LHC Project Report 691 (2004).

[Bal12] BALLARINO, A: *Operation of 1074 HTS current leads at the LHC machine: overview of three years of powering the accelerator*. In: IEEE Transactions on Applied Superconductivity 23 (2013), Nr. 3, Pag. 4801904.

[Bar13] BARTH, C.: *High temperature superconductor cable concepts for fusion magnets*. Dissertation, KIT Scientific Publishing (2013).

[Bau12] BAUER, P.: Private communication. ITER Organisation. 2012.

[BBB12] BALLARINO, A.; BAUER, P.; BI, Y.; DEVRED, A.; DING, K.; FOUSSAT, A; MITCHELL, N.; SHEN, G.; SONG, Y.; TAYLOR, T; YANG, Y; ZHOU, T.: *Design of the HTS current leads for ITER*. In: IEEE Transactions on Applied Superconductivity 22 (2012), Nr. 3, Pag. 4800304.

[BC76] BEJAN, A.; CLUSS, Jr. E. M.: *Criterion for burn-up conditions in gas-cooled cryogenic current leads*. In: Cryogenics 16 (1976), Pag. 515-518.

[Bej84] BEJAN, A.: *Convection heat transfer*. Wiley and Sons, 1984.

[BFS75] BUYANOV, Y. L.; FRADKOV, A. B.; I. Y. SHEBALIN: *A review of current leads for cryogenic devices*. In: Cryogenics 15 (1975), Pag. 193-200.

[BH13] BRADSHAW, A. M.; HAMACHER, T.: *Nuclear fusion and the helium supply problem*. In: Fusion Engineering and Design (2013).

[BMM03] BALLARINO, A; MATHOT, S; MILANI, D.: *13000 A HTS current leads for the LHC accelerator: from conceptual design to prototype validation*. In: Proceedings of EUCAS (2003), LHC Project Report 696.

[Buc14] BUCKINGHAM, E.: *On physically similar systems; Illustration of the use of dimensional equations*. In: Physical Review 4 (1914), Pag. 345-376.

[Buy85] BUYANOV, Y. L.: *Current leads for use in cryogenic devices. Principle of design and formulae for design calculations*. In: Cryogenics 25 (1985), Pag. 94-110.

[CVS98] CHANG, H.-M.; VAN SCIVER S. W.: *Thermodynamic optimization of conduction-cooled HTS current leads*. In: Cryogenics 38 (1998), Nr. 7, Pag. 729-736.

[Cryo] http://cryogenics.nist.gov/MPropsMAY/304Stainless/304Stainless_rev.htm (July, 1st 2013).

[Czy04] CZYCHOLL, G.: *Theoretische Festkörperphysik. 2nd Auflage*. Springer-Verlag Berlin Heidelberg (2004).

[Eki78] EKIN, J. W.: *Current transfer in multifilamentary superconductors. I. Theory*. In: Journal of Applied Physics 49 (1978), Nr. 6, Pag. 3406.

[EU07] *Agreement between the European atomic energy community and the government of Japan for the joint implementation of the broader approach activities in the field of fusion energy research*. In: Official Journal of the European Union, September, 21th 2007.

[EU10] EUROPEAN COMMISSION: *ITER: present status and future perspectives*. Brüssel, May 4th 2010.

[ETP12] *Energy technology perspective 2012 – Executive summary*. International Energy Agengy (2012).

[FDF11] FIETZ, W. H.; DROTZIGER, S.; FINK, S.; HEIDUK, M.; HELLER, R.; KOPMANN, A.; LANGE, C.; LIETZOW, R.; MOHRING, T.; ROHR, P.; RUMMEL, T.; SÜSSER, M.: *Test Arrangement for the W7-X HTS-Current Lead Prototype Testing*. In: IEEE Transactions on Applied Superconductivity 21 (2011), Nr. 3, Pag. 1058-1061.

[FHK09] FIETZ, W. H.; HELLER, R.; KIENZLER, A.; LIETZOW, R.: *High Temperature Superconductor Current Leads for WENDELSTEIN 7-X and JT-60SA*. In: IEEE Transactions on Applied Superconductivity 19 (2009), Nr. 3, Pag. 2202-2205.

[Fic72] FICKETT, F. R.: /. In: Int. Copper Res. Rep. 186 (1972).

[FZK04] HELLER, R.; DARWESCHSAD, S. M.; DITTRICH, G.; FIETZ, W. H.; FINK, S.; HERZ, W.; KIENZLER, A.; LINGOR, A.; MEYER, I.; NÖTHER, G.; SÜSSER, M.; TANNA, V. L.; WESCHE, R.; WÜCHNER, F.; ZAHN, G.: *Final report on the test of the 70 kA HTS current lead in the TOSKA facility*. Forschungszentrum Karlsruhe interner Bericht (2004).

[FZK05] HELLER, R.; BORLEIN, M.; DARWESCHSAD, S. M.; DITTRICH, G.; FIETZ, W. H.; FINK, S.; HERZ, W.; KIENZLER, A.; LINGOR, A.; MEYER, I.; SÜSSER, M.; TANNA, V. L.; WESCHE, R.; WÜCHNER, F.; ZAHN, G.: *Report of the phase III operation of the 70 kA HTS current lead*. Forschungszentrum Karlsruhe interner Bericht (2005).

[GMP04] GONZÁLEZ, J.; MESTRES, N.; PUIG, T.; GÁZQUEZ, J.; SANDIUMENGE, F.; OBRADORS, X.; USOSKIN, A.; JOOSS, C.; FREYHARDT, H.; FEENSTRA, R.: *Biaxial texture analysis of YBa2Cu3O7-x coated conductors by micro-Raman spectroscopy*. In: Physical Review B 70 (2004), Nr. 9, Pag. 1–8.

[GSR08] GOLDACKER, W.; SCHLACHTER, S. I.; RINGSDORF, B.; SCHMIDT, C.; WEISS, K-P.; SCHWARZ, M.; FRANK, A.; LAMPE, A.; KLING, A.; HELLER, R.: *Final report on HTS material for fusion magnets*: EFDA-No. TW5-TMS-HTSPER /
Forschungszentrum Karlsruhe GmbH (FZK). Eggenstein-Leopoldshafen, 2008. –Forschungsbericht.

[GWB12] GEIGER, J.; WOLF, R. C.; BEIDLER, C.; CARDELLA, A.; CHLECHOWITZ, E.; ERCKMANN, V.; GANTENBEIN, G.; HATHIRAMANI, D.; HIRSCH, M.; KASPAREK, W.; KISSLINGER, J.; KOENIG, R.; KORNEJEW, P.; LAQUA, H. P.; LECHTE, C.; LORE, J.; LUMSDAINE, A.; MAASSBERG, H.; MARUSHCHENKO, N. B.; MICHEL, G.; OTTE, M.; PEACOCK, A.; SUNN PEDERSEN, T.;

THUMM, M.; TURKIN, Y.; WERNER, A.; ZHANG, D.; the W7-X Team: *Aspects of steady-state operation of the Wendelstein 7-X stellarator*. In: Plasma Physics and Controlled Fusion 55 (2013), 014006.

[HAA04] HELLER, R.; AIZED, D.; AKHMETOV, A.; FIETZ, W. H.; HURD, F.; KELLERS, J.; KIENZLER, A.; LINGOR, A.; MAGUIRE, J.; VOSTNER, A.; WESCHE, R.: *Design and fabrication of a 70 kA current lead using Ag/Au stabilized Bi-2223 tapes as a demonstrator for the ITER TF-coil system*. In: IEEE Transactions on Applied Superconductivity 14 (2004), Nr. 2, Pag. 1774-1777.

[Han79] HANDS, B. A.: *HEPROP – a computer programme for the thermodynamic and thermophysical properties of helium – Third Edition*. In: Report of the Department of Engineering Science, University of Oxford (1979).

[HC77] HO, C. Y.; CHU, T. K.: *Electrical resistivity and thermal conductivity of nine selected AISI stainless steels*. In: CINDAS Report 45, September 1977.

[HCB10] HELLER, R.; CLASS, A.; BATTA, A.; LIETZOW, R.; NEUMANN, H.; TISCHMACHER, M.: *Modelling of the fin type heat exchanger for the HTS current leads of W7-X and JT-60SA*. In: Cryogenics 50 (2010), Nr. 3, Pag. 222-230.

[HDD05] HELLER, R.; DARWESCHSAD, S. M.; DITTRICH, G.; FIETZ, W. H.; FINK, S.; HERZ, W.; HURD, F.; KIENZLER, A.; LINGOR, A.; MEYER, I.; NOTHER, G.; SÜSSER, M.; TANNA, V. L.; VOSTNER, A.; WESCHE, R.; WUCHNER, F.; ZAHN. G.: *Experimental results of a 70 kA high temperature superconductor current lead demonstrator for the ITER magnet system*. In: IEEE Transactions on Applied Superconductivity 15 (2005), Nr. 2, Pag. 1496-1499.

[HDF11a] HELLER, R.; DROTZIGER, S.; FIETZ, W. H.; FINK, S.; HEIDUK, M.; KIENZLER, A.; LANGE, C.; LIETZOW, R.; MOHRING, T.; ROHR, P.; RUMMEL, T.; MONNICH, T.; BUSCHER, K.-P.: *Test Results of the High Temperature Superconductor Prototype Current Leads for*

Wendelstein 7-X. In: IEEE Transactions on Applied Superconductivity 21 (2011), Nr. 3, Pag. 1062-1065.

[HDF11b] HELLER, R.; DROTZIGER, S.; FIETZ, W. H.; FINK, S.; HEIDUK, M; LANGE, C.; LIETZOW, R.; MOHRING, T.; ROHR, P.; RUMMEL, T.; NAGEL, M.: *Acceptance of results of W7-X HTS current leads prototype test*.
W7X-CQM-SZFP1P2-PX-0002 (2011).

[Hel89] HELLER, R.: *Numerical calculation of current leads for fusion magnets*. In: KfK (Kernfonrschungszentrum Karlsruhe) (1989), Nr. 4608, Karlsruhe.

[Hel09] HELLER, R.: *Safety Analysis of the 70 kA ITER HTS Current Lead Demonstrator*. In: IEEE Transactions on Applied Superconductivity 19 (2009), Nr. 3, Pag. 1504-1507.

[Hel13a] HELLER, R.: *Private communication about the LOFA at I = 18.2 kA in the W7-X HTS current leads*. 2013.

[Hel13b] HELLER, R.: *Private communication about measured data implemented in the code CURLEAD*. 2013.

[HFK03] HELLER, R.; FRIESINGER, G.; KIENZLER, A.; KOMAREK, P.; LINGOR, A.; ULBRICHT, A.; ZAHN, G.: *Construction and operation performance of the 80 kA current leads used for the test of the ITER toroidal field model coil in the TOSKA facility*. In: IEEE Transactions on Applied Superconductivity 13 (2003), Nr. 2,
Pag. 1918-1921.

[HFK08] HELLER, R.; FIETZ, W. H.; KELLER, P.; RINGSDORF, B.; SCHLACHTER, S. I.; SCHWARZ, M.; WEISS, K. P.: *Electrical, mechanical and thermal characterization of Bi-2223/AgAu material for use in HTS current leads for W7-X*. In: IEEE Transactions on Applied Superconductivity 18 (2008), Nr. 2, Pag. 1443-1446.

[HFK11] HELLER, R.; FIETZ, W. H.; KIENZLER, A.; LIETZOW, R.: *High temperature superconductor current leads for fusion machines*. In: Fusion Engineering and Design 86 (2011), Nr. 6-8, Pag. 1422-1426.

[HFL06] HELLER, R.; FIETZ, W. H.; LIETZOW, R.; TANNA, V. L.; VOSTNER, A.; WESCHE, R.; ZAHN, G. R.: *70 kA High Temperature Superconductor Current Lead Operation at 80 K*. In: IEEE Transactions on Applied Superconductivity 16 (2006), Nr. 2, Pag. 823-826.

[Hil77] HILAL, M.: *Optimization of current leads for superconducting systems*. In: IEEE Transactions on Magnetics 13 (1977), Nr. 1, Pag. 690-693.

[HKS99] HOBL, A.; KRISCHEL, D.; SCHILL, M.; SCHAEFER, P.: *HTc current leads in commercial magnet system applying Bi2212 MCP BSCCO material*. In: IEEE Transaction on Applied Superconductivity 9 (1999), Nr. 2.

[HL07] HELLER, R.; LIETZOW, R.: *Results of the heat exchanger mock-up tests*. Interner Bericht FE.5130.0086.0014/A. Forschungszentrum Kalrsurhe, Institut für Technische Physik, Karlsruhe, Germany.

[Hul03] HULL, J. R.: *Application of high temperature superconductors in power technology*. In: Reports on Progress in Physics 66 (2003), Nr. 11.

[IBK10] ISHIDA, S.; BARABASCHI, P.; KAMADA, Y.: *Status and prospect of the JT-60SA project*. In: Fusion Engineering and Design 85 (2010), Nr. 10-12, Pag. 2070-2079.

[IBK11] ISHIDA, S.; BARABASCHI, P.; KAMADA, Y.; JT-60SA TEAM: *Overview of the JT-60SA project*. In: Nuclear Fusion 51 (2011), Pag. 094018.

[Ide86] IDELCHIK, I.: *Handbook of hydraulic resistance*. Hemisphere Publishing Corporation, 1986.

[ITEa] http://www.iter.org/proj/iterhistory
(June, 17th 2013).

[Joh61] JOHNSON, V.J.: *Properties of Materials at Low Temperature (Phase 1)*. A Compendium. Pergamon Press, 1961.

[JT60] JT-60SA RESEARCH UNIT: *JT-60 SA Research plan*.
http://www.jt60sa.org/pdfs/JT-60SA_Res_Plan.pdf
(July, 1st 2013).

[Kit96] KITTEL, C.: *Introduction to solid state physics*. Wiley and Sons, 1996.

[KNI98] KOIZUMI, K.; NAKAHIRA, Y.; ITOU, Y.; TADA, E.; JOHNSON, G.; IOKI, K.; ELIO, F.; IIZUMA, T.; SANNAZZARO, G.; TAKAHASHI, K.; UTIN, Y.; ONOZUKA, M.; NELSON, B.; VALLONE, C.; KUZMIN, E.: *Design and development of the ITER vacuum vessel.* In: Fusion Engineering and Design 41 (1998), Pag. 299-304.

[Kom95] KOMAREK, P.: *Hochstromanwendungen der Supraleitung.* 1. B.G. Teubner Stuttgart, 1995.

[KSA87] KAKAC, S.; SHAH, R. K.; AUNG, W.: *Handbook of single-phase convective heat transfer.* John Wiley & Sons, 1987.

[KSW09] KELLER, P.; SCHWARZ, M.; WEISS, K.-P.; HELLER, R.; JUNG, A.; AUBELE, A.: *Electromechanical and thermal characterization of stacked Bi-2223 tapes at cryogenic temperature.* In: IEEE Transactions on Applied Superconductivity 19 (2009), Nr. 3, Pag. 2893-2896.

[LDR99] LE BARS, J.; DECHAMBRE, T.; ROGUIER, P.; GAGNANT, K.: *Development of current leads using electronically deposited BSCCO 2212 tapes.* In: IEEE Transaction on Applied Superconductivity 9 (1999), Nr. 2.

[LHN08] LIETZOW, R.; HELLER, R.; NEUMANN, H.: *Performance of heat exchanger models in up-side-down orientation for the use in HTS current leads for W7-X.* In: AIP Conference Proceedings 985 (2008), Nr. 1, Pag. 1243-1250.

[MDL12] MITCHELL, N.; DEVRED, A.; LIBEYRE, P.; LIM, B.; SAVARY, F.: *The ITER magnets: design and construction status.* In: IEEE Transaction on Applied Superconductivity 22 (2012), Nr. 3, Pag. 4200809.

[Men94] MENTER, F. R.: *Two-equation eddy-viscosity turbulence models for engineering applications.* In: AIAA Journal 32 (1994), Nr. 8, Pag. 1598-1604.

[MLV06] MENTER, F. R.; LANGTRY, R.; VÖLKER, S.: *Transition modelling for general purpose CFD codes.* In: Flow Turbulence Combust 77 (2006), Pag. 277-303.

[MMM03] MARKEN, K. R.; MIAO, H.; MEINESZ, M.; CZABAJ, B.; HONG, S.: *BSCCO-2212 conductor development at oxford superconducting technology*. In: IEEE Transactions on Applied Superconductivity 13 (2003), Nr. 2, Pag. 3335–3338.

[MRV89] MATRONE, A.; ROSATELLI, G.; VACCARONE, G.: *Current leads with high Tc superconducting bus bars*. In: IEEE Transactions on Magnetics 25 (1989), Nr. 2, Pag. 1742-1745.

[MTF88] MAEDA, H.; TANAKA, Y.; FUKUTUMI, M.; ASANO, T.: *A New high-Tc oxide superconductor without a rare earth element*. In: Japanese Journal of Applied Physics 27 (1988), Nr. 2, Pag. 209-207.

[Mum89] MUMFORD, F. J.: *Superconducting current-leads made from high Tc superconductor and normal metal conductor*. In: Cryogenics (1989), Nr. 29, Pag.206-207.

[MYI07] MUROYAMA, S.; YAGI, M.; ICHIKAWA, W.; TORII, S.; TAKAHASHI, T.; SUZUKI, H.; YASUDA, K.: *Experimental results of a 500 m HTS power cable field test*. In: Transaction on Applied Superconductivity 17 (2007), Nr. 2, Pag. 1680-1683.

[Nex] http://www.nexans.de/eservice/Germany-en/navigate_299973/Components.html (June, 12th 2013).

[Oak96] OAK RIDGE NATIONAL LABORATORIES: *The chemistry of superconductors*.
In: http://www.ornl.gov/info/reports/m/ornlm3063r1/pt5.html. (1996). (June, 13th 2013).

[PKL09] PITTS, R. A.; KUKUSHKIN, A.; LOARTE, A.; MARTIN, A.; MEROLA, M.; KESSEL, C. E.; KOMAROV, V.; SHIMADA, M.: *Status and physics basis of the ITER divertor*. In: Physica Scripta (2009), 014001.

[PS72] PATANKAR, S. V.; SPALDING, D. B.: *A calculation procedure for heat, mass and momentum transfer in three-dimensional parabolic flows*. In: International Journal of Heat and Mass Transfer 15 (1972), Nr. 10, Pag. 1787-1806.

[Ras11] RASMUSSEN, D.: *Overview of ITER heating and current drive systems.* Presented at: APS-DPP USBPO ITER Town Meeting (November 5th, 2011).

[RHS11] RIZZO, E.; HELLER, R.; SAVOLDI RICHARD, L.; ZANINO, R.: *Heat exchanger CFD analysis for the W7-X high temperature superconductor current lead prototype.* In: Fusion Engineering and Design 86 (2011), Nr. 6-8, Pag. 1571-1574.

[RHS13a] RIZZO, E.; HELLER, R.; SAVOLDI RICHARD, L.; ZANINO, R.: *CtFD-based correlations for the thermal–hydraulics of an HTS current lead meander-flow heat exchanger in turbulent flow.* In: Cryogenics 53 (2013), Pag. 51-60.

[RHS13b] RIZZO, E.; HELLER, R.; SAVOLDI RICHARD, L.; ZANINO, R.: *Computational thermal-Fluid Dynamics analysis of the laminar regime in the meander flow geometry characterizing the heat exchanger used in the High Temperature Superconducting current leads.* In: Fusion Engineering and Design 88 (2013), Nr. 11, Pag. 2749-2756.

[RHS13c] RIZZO, E.; HELLER, R.; SAVOLDI RICHARD, L.; ZANINO, R.: *1-D thermal-hydraulic analysis of the high temperature superconducting current leads for the ITER magnet system from 5 K to 300 K.* In: Fusion Engineering and Design 88 (2013), Nr. 12, Pag. 3125-3131.

[RM11] RAFFRAY, A. R.; MEROLA, M.: *Design of the ITER first wall and blanket.* In: 2011 IEEE/NPSS 24th Symposium on Fusion Engineering (2011).

[RRE11] RUMMEL, T.; RISSE, K.; EHRKE, G.; RUMMEL, K.; JOHN, A.; MONNICH, T.; BUSCHER, K.-P.; FIETZ, W. H.; HELLER, R.; NEUBAUER, O.; PANIN, A.: *The superconducting magnet system of the stellarator Wendelstein 7-X.* In: Fusion Engineering (2011), Pag. 1-6.

[RWH92] ROSNER, C. H.; WALKER, M. S.; HALDAR, P.; MOTOWIDLO, L. R.: *Status of HTS superconductors: Progress in improving transport critical current densities in HTS Bi-2223 tapes and coils.* In: Cryogenics 32 (1992), Nr. 11, Pag. 940-948.

[SBB13] SITKO, M.; BORDINI, B.; BALLARINO, A.; BAUER, P.; DEVRED, A.: *3D Numerical Analysis of the 68 kA Heat Exchanger for ITER TF Current Leads*. In: IEEE Transactions on Applied Superconductivity 23 (2013), Nr. 3, Pag. 4900805.

[SCF10] SAVOLDI RICHARD, L.; CLASS, A.; FIETZ, W. H.; HELLER, R.; RIZZO, E.; ZANINO, R.: *CtFD Analysis of HTS Current Lead Fin-Type Heat Exchanger for Fusion Applications*. In: IEEE Transactions on Applied Superconductivity 20 (2010), Nr. 3, Pag. 1733-1736.

[Sch09] SCHWARZ, M.: *Wärmeleitfähigkeit supraleitender Kompositleiter im Temperaturbereich von 4 K bis 300 K*. Dissertation, 2009.

[Sel11] SELVAMANICKAM, V.: *Second-generation HTS Wire for Wind Energy Applications*. In: Symposium on Superconducting Devices for Wind Energy. Barcelona, (2011).

[Sip04] SIPS, A. C. C.: *Advanced scenarios for ITER operation*. In: Plasma Physics and Controlled Fusion 47 (2005), Pag. A19-A40.

[SKL07] STENVALL, A.; KORPELA, A; LEHTONEN, J.; MIKKONEN, R.: *Current transfer length revisited*. In: Superconductor Science and Technology 20 (2007), Nr. 1, Pag. 92-99.

[Sta08] CD-ADAPCO: Methodology, Star-CD version 4.06. 2008.

[StaCC] CD-ADAPCO: Online user manual, Star-CCM+ version 6.02.009.

[SWH09] SCHWARZ, M.; WEISS, K. P.; HELLER, R.; FIETZ, W. H.: *Thermal conductivity measurement of HTS tapes and stacks for current lead applications*. In: Fusion Engineering and Design 84 (2009), Nr. 7-11, Pag. 1748-1750.

[Tan06] TANNA, V. L.: *Design and analysis of the superconducting current feeder system for the International Thermonuclear Reactor (ITER)*. Dissertation, Forschungszentrum Karlsruhe Wissenschaftliche Berichte FZKA 7256.

[Tay12] TAYLOR, T.: *Design report of the 55 kA and 68 kA HTS current leads*. In: Internal Report ITER_D_3QNTXG. 2012.

[TMC97] TING, S. M.; MARKEN, K. R. Jr., COWEY, L.; DAI, W.; HONG, S.; NELSON, S.: *Development of current leads using dip coated BSCCO-2212 tape*. In: IEEE Transaction on Applied Superconductivity 7 (1997), Nr. 2.

[W7X] Wendelstein 7-X: Research status. http://www.ipp.mpg.de/ippcms/eng/pr/forschung/w7x/stand/index.html (May, 31st 2013).

[Wan00] WANNER, M.; the W7-X Team: *Design goals and status of the WENDELSTEIN 7-X project*. In: Plasma Physics and Controlled Fusion 42 (2000), Pag. 1179-1186.

[WEF03] WANNER, M.; ERCKMANN, V.; FEIST, J.-H.; GERDEBRECHT, W.; HARTMANN, D.; KRAMPITZ, R.; NIEDERMEYER, H.; RENNER, H.; RUMMEL, T.; SCHAUER, F.; WEGENER, L.; WESNER, F.; MUELLER, G. A.; KASPAREK, W.; THUMM, M.; DAMMERTZ, G.: *Status of WENDELSTEIN 7-X construction*. In: Nuclear Fusion 43 (2003), Pag. 416-424.

[Weg09] WEGENER, L.: *Status of Wendelstein 7-X construction*. In: Fusion Engineering and Design 84 (2009), Nr. 2-6, Pag. 106-112.

[WEO12] *World Energy Outlook 2012*. International Energy Agency, 2012.

[WF1853] WIEDEMANN, G.; FRANZ. R.: *Ueber die Wärme-Leitungsfähigkeit derMetalle*. In: Annalen der Physik. 165 (1853), Nr. 8, Pag. 497-531.

[Wika] http://en.wikipedia.org/wiki/BSCCO (June, 13th 2013).

[Wil94] WILCOX, D. C.: *Simulation of transition with a two-equation model*. In: AIAA Journal 32 (1994), Nr. 2, Pag. 247-255.

[Wil98] WILCOX, D. C.: *Turbulence modeling for CFD*. DCW Industries, Inc., 1998.

[Wils83] WILSON, M. N.: *Superconducting magnets*. Oxford University Press, 1983.

[WSL07] WEISS, K. P.; SCHWARZ, M.; LAMPE, A.; HELLER, H.; FIETZ, W. M.; NYILAS, A.; SCHLACHTER, S.; GOLDHACKER, W.:

Electromechanical and thermal properties of Bi2223 tapes. In: IEEE Transactions on Applied Superconductivity 17 (2007), Nr. 2, Pag. 2079-2082.

[YTK10] YOSHIDA, K.; TSUCHIYA, K.; KIZU, K.; MURAKAMI, H.; KAMIYA, K.; OBANA, T.; TAKAHATA, K.; PEYROT, M.; BARABASCHI, P.: *Development of JT-60SA superconducting magnet system*. In: Physica C 470 (2010), Pag. 1727-1733.

[ZGM06] ZANINO, R.; GIORS, S.; MONDINO, R.: *CFD modeling of ITER Cable-in-Conduit Superconductors. Part I : friction in the central channel*. In: Advances in Cryogenic Engineering 51 (2006), Pag. 1009–1016.

Karlsruher Schriftenreihe zur Supraleitung (ISSN 1869-1765)

Herausgeber: Prof. Dr.-Ing. M. Noe, Prof. Dr. rer. nat. M. Siegel

Die Bände sind unter www.ksp.kit.edu als PDF frei verfügbar oder als Druckausgabe bestellbar.

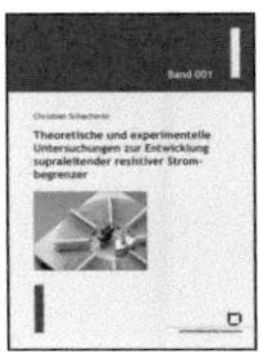

Band 001
Christian Schacherer
Theoretische und experimentelle Untersuchungen zur Entwicklung supraleitender resistiver Strombegrenzer. 2009
ISBN 978-3-86644-412-6

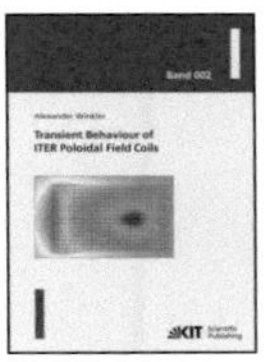

Band 002
Alexander Winkler
Transient behaviour of ITER poloidal field coils. 2011
ISBN 978-3-86644-595-6

Band 003
André Berger
Entwicklung supraleitender, strombegrenzender Transformatoren. 2011
ISBN 978-3-86644-637-3

Band 004
Christoph Kaiser
High quality Nb/Al-AlOx/Nb Josephson junctions. Technological development and macroscopic quantum experiments. 2011
ISBN 978-3-86644-651-9

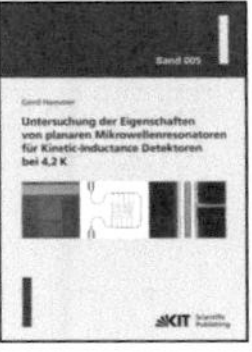

Band 005
Gerd Hammer
Untersuchung der Eigenschaften von planaren Mikrowellen-resonatoren für Kinetic-Inductance Detektoren bei 4,2 K. 2011
ISBN 978-3-86644-715-8

Band 006
Olaf Mäder
Simulationen und Experimente zum Stabilitätsverhalten von HTSL-Bandleitern. 2012
ISBN 978-3-86644-868-1

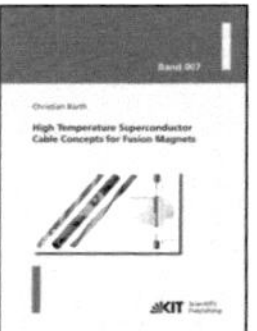

Band 007
Christian Barth
High Temperature Superconductor Cable Concepts for Fusion Magnets. 2013
ISBN 978-3-7315-0065-0

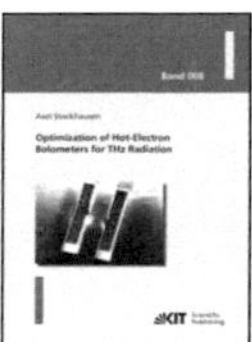

Band 008
Axel Stockhausen
Optimization of Hot-Electron Bolometers for THz Radiation. 2013
ISBN 978-3-7315-0066-7

Band 009
Petra Thoma
Ultra-fast $YBa_2Cu_3O_{7-x}$ direct detectors for the THz frequency range. 2013
ISBN 978-3-7315-0070-4

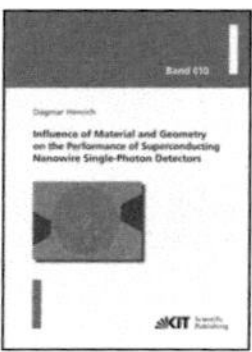

Band 010
Dagmar Henrich
Influence of Material and Geometry on the Performance of Superconducting Nanowire Single-Photon Detectors. 2013
ISBN 978-3-7315-0092-6

Band 011
Alexander Scheuring
Ultrabreitbandige Strahlungseinkopplung in THz-Detektoren. 2013
ISBN 978-3-7315-0102-2

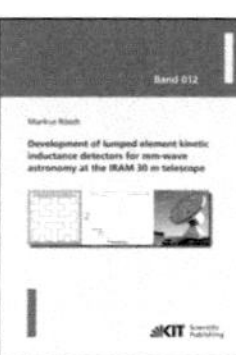

Band 012
Markus Rösch
Development of lumped element kinetic inductance detectors for mm-wave astronomy at the IRAM 30 m telescope. 2013
ISBN 978-3-7315-0110-7

Band 013
Johannes Maximilian Meckbach
Superconducting Multilayer Technology for Josephson Devices. 2013
ISBN 978-3-7315-0122-0

Band 014
Enrico Rizzo
Simulations for the optimization of High Temperature Superconductor current leads for nuclear fusion applications. 2014
ISBN 978-3-7315-0132-9